Basic Numeracy Skills and Practice

Basic Numeracy Skills and Practice

J. Newbury, PhD

M

First published 1981 by
THE MACMILLAN PRESS LTD
London and Basingstoke
Companies and representatives throughout the world

ISBN 0 333 29336 3

Printed in Hong Kong

Typeset in 10/12 Century by Illustrated Arts

Contents

Preface

This text presents the basic numeracy skills required in an introductory course in mathematics. The skills are presented through a series of logical steps that will establish confidence in attempting any of the questions. The text is suitable for children and teachers in primary schools, for pupils starting a course in CSE mathematics at the various modes, and for the adult returning to study after a number of years. It is suitable for self-help group teaching or for individuals working by themselves.

The text may be used as a primary teaching book or supplementary material to a main textbook. To facilitate ease of working, exercises are graded within each group from elementary to more complex since I believe proficiency in mathematics is directly related to practice.

The text commences with an introduction to the way problems may be attempted and progresses through fractions, numbers, transposition, equations, graphical methods and their applications, to indices, logarithms and finally an introduction to angles and trigonometrical ratios.

Each chapter presents the theory and principles which are then applied through worked examples. At the end of each section additional questions are presented. Revision exercises are found at intermediate parts of the book — the answers at the end of the book are amplified where necessary, especially in the case of graphical questions. The text has been presented such that the student may study progressively or use individual sections without loss of clarity.

The combination of presenting the principles and concepts of elementary numeracy through worked examples and additional graded questions will I trust give confidence to the reader. The questions, exercises and solutions are completely my responsibility.

I thank colleagues and students who have given advice whilst the manuscript has been in preparation. To John Watson and Peter Milford of Macmillan Press my special thanks throughout the preparation and editorial stages.

John Newbury 1980

1

Introduction to Mathematics

Why is mathematics a useful activity to engage in, and what may we reasonably expect to obtain in return for the effort of learning even more mathematics?

Historically speaking, mathematics developed as a language for describing problems and as a tool for solving these problems. At a later stage people began to be interested in the nature of the tools they had invented and thus an interest in 'mathematics for its own sake' was born.

You have already begun to imitate this development; very early in childhood you used the numerals 1, 2, 3 . . . as convenient labels for ideas of number — the first steps in acquiring some mathematical language. Later on you began to enlarge this language, introducing addition and so on, and to manipulate numbers to solve simple problems of a type which contained the words 'how many . . . ?'

'How many?' is already half a step removed from the pure language of number, because this question implies that we are somehow involved with real objects which really matter — perhaps we are buying or selling them. The next question 'how much?' indicates even more — we have a problem of quantity, concerning land, money, cloth, food or other goods — and it implies that we have some feasible way of measuring the quantities involved.

Once we can measure things we *must* use numbers, and there is a good chance that mathematics can help. If we have a problem of quantity, we may try to convert it into a problem of mathematics (the simpler the better of course). If we succeed, possibly not the whole, but an essential part of the problem in real life is replaced by a mathematical problem.

We have now a 'mathematical model' of the real-life problem.

What sort of problems may we try to solve by mathematical means? What type of answers may we expect to get? There is a temptation to say that only problems asking 'how many?' or 'how much?' are suitable for mathematical solution; such a view would be much too restrictive and there is probably no satisfactory answer to the question. One possible partial answer is that any situation in which we suspect that one event causes another, or in which two types of behaviour are related, is suitable for mathematical investigation.

How do we apply mathematical ideas to a problem? Any real problem is unique and will require individual treatment, but there are certain recognisable steps in attempting to solve most problems. We will list the steps here with brief comments.

(1) Understand the problem — Most real (as opposed to textbook) problems are not stated carefully; the question asked may be vague; too little (or too much) information may be available.

(2) Decide specific questions

(3) Decide what is relevant — There may be much irrelevant information given as well as useful information. Ideas on what is relevant may have to be changed later on.

(4) Describe the problem mathematically — At this point it may be necessary to invent some extra mathematical language or notation.

(5) Solve the mathematical problem — This step may involve solving a number of smaller mathematical problems on the way.

(6) Interpret the mathe-   That is, what does it mean in
matical solution   real terms?

(7) Test your conclusions   Steps 3−6 will probably have
                             involved assumptions and
                             simplifications; it is wise to
                             test some conclusions by
                             experiment if it is at all
                             practicable.

(8) If necessary return to
step 3 and alter your
assumptions.

In most textbook problems steps 1, 2, 3, 7 and 8 are ignored;
all that is asked is that you do part of step 4, step 5 and part
of step 6. Unfortunately the steps left out are the ones which
are absolutely vital to any serious application of mathematics
because they are the ones which involve the relationship
between mathematics and the real world.

First of all let me specify the things which you are expected to know before starting to study the contents of this
book. I shall assume very little indeed:

(1) The ability to carry out the basic arithmetical operations
of *addition*, *subtraction*, *multiplication* and *division* using
whole numbers.

Examples: I assume that you can evaluate 22 + 99,
123 − 45, 24 x 68 and 923 ÷ 13, and that you will agree
with the answers of 121, 78, 1632 and 71 respectively.

(2) The ability to *recognise* the need for those numbers
which are usually known as *negative numbers*.

Examples: If I wish to indicate that 3 people share
a cake equally among them, then each of them
gets one-third of the cake; if we are thirsty, but not
very thirsty, then we may order a half-pint of beer
(indicated by the fraction ½); if I am measuring a pane
of glass, then I have to specify its size as 21½ in by 9⅞

in, say. If I wish to speak of a place which is 100 feet below sea level then I may describe its 'height' as −100 feet; a very cold night may have resulted in a recorded temperature of −6 °C; and if I find myself temporarily embarrassed by being overdrawn at the bank I say that I am £10 'in the red', or that I am £10 'under', and I could describe this by saying that my balance is −£10.

I make *no* assumptions as to how to *manipulate* such numbers, that is, I do *not* assume that you can evaluate ¾ + ⅝, or (−2) x (−5), these I shall explain as we proceed. In brief I am assuming that you can *do arithmetic* using positive whole numbers, and that you can *see the need* for fractions and negative numbers to describe some of the facts of everyday life.

So first I shall talk about fractions and find a way to do arithmetic using such numbers. Then I shall tackle the negative numbers in a similar fashion.

2

Fractions

We know that fractions are necessary in measurement so let us start there. Suppose you wanted to know the length of the line ABC and you were told that AB is 2¾ in long and BC is 1½ in long. The problem is how do you add 2¾ to 1½?

A   B   C

There are several ways of doing this, one of which I will use here. The numbers 2¾ and 1½ are called *mixed numbers* because they are a mixture between whole numbers, 2 and 1, and fractions, ¾ and ½. So let me start by converting the mixed numbers into fractions, first of all, 2¾.

How many ¼s are there in 2? There are 8, so 2¾ expressed in quarters is 8/4 + 3/4 = 11/4. In the same way, converting 1½ into a fraction gives 3/2, that is there are 3 halves in 1½. Our problem now is, what is 11/4 + 3/2 ?

If our problem had been, what is 11/4 + 5/4 then the answer would have been 16/4. (That is, 11 quarters plus 5 quarters equals 16 quarters.) So a way for us to solve our problem is to convert all the fractions to the same kind, in this case quarters. So

$$\frac{11}{4} + \frac{3}{2} = \frac{11}{4} + \frac{6}{4} = \frac{11+6}{4} = \frac{17}{4}$$

Notice what happened, we changed the 2 into a 4 by multiplying it by 2, and we also changed the 3 into a 6 by multiplying it by 2, that is, we scaled the top number by the same amount as we scaled the bottom number. In mathematical language the top number is called the *numerator* and the bottom number the *denominator*. I can express what we did in another way, that is, we converted all our fractions into fractions with the *same denominator*. We also scaled a fraction in order to obtain an equivalent form. So

$$\frac{3}{13} = \frac{6}{26} = \frac{9}{39} \text{ etc.}$$

and

$$\frac{1}{4} = \frac{2}{8} = \frac{4}{16} = \frac{8}{32} \text{ etc.}$$

Example: Evaluate $\frac{3}{12} + 1\frac{1}{3}$

$$\frac{3}{12} + 1\frac{1}{3} = \frac{3}{12} + \frac{4}{3}$$

We now have only *fractions* but the denominators are different. The lowest common denominator here is 12.

Therefore

$$= \frac{3}{12} + \frac{16}{12}$$

$$= \frac{19}{12}$$

$$= 1\frac{7}{12}$$

Exercise 1: Evaluate $\frac{5}{6} + \frac{2}{7}$

Exercise 2: Evaluate $\frac{2}{3} + \frac{5}{11}$

Exercise 3: Evaluate $6\frac{2}{3} + 1\frac{1}{2}$

The *subtraction* of fractions uses the same idea of the *same* denominator or *common denominator* as it is usually called.

Example: Evaluate $3\frac{1}{2} - 2\frac{1}{3}$

$$3\frac{1}{2} - 2\frac{1}{3} = \frac{7}{2} - \frac{7}{3}$$

$$= \frac{21}{6} - \frac{14}{6}$$

$$= \frac{21 - 14}{6}$$

$$= \frac{7}{6} = 1\frac{1}{6}$$

Exercise 4: Evaluate $2\frac{1}{4} - \frac{2}{3}$

Exercise 5: Evaluate $2 - \frac{1}{4}$

Exercise 6: Evaluate $\frac{2}{3} - \frac{3}{5}$

Now how about *multiplying* fractions? What is 3 x 1½? I am sure you will say 4½ no matter how you did it but we can follow a similar method as we have used for the addition of fractions. So

$$3 \times 1\frac{1}{2} = \frac{3}{1} \times \frac{3}{2}$$

$$= \frac{6}{2} \times \frac{3}{2}$$

$$= ? \qquad \text{Is it } \frac{18}{4} \text{ or } \frac{6 \times 3}{2} = \frac{18}{2}?$$

The rule for multiplying fractions is simply this — multiply the top numbers, multiply the bottom numbers, and divide. Of course you must always begin by expressing the mixed numbers like 1½ in purely fractional form like 3/2. So the correct answer is 18/4 or 4½.

Example: $1\frac{1}{2} \times 2\frac{1}{2} \times 3\frac{1}{3}$ $= \frac{3}{2} \times \frac{5}{2} \times \frac{10}{3}$

$$= \frac{150}{12} = \frac{25}{2} = 12\frac{1}{2}$$

Exercise 7: Evaluate $1\frac{1}{2} \times 1\frac{1}{3} \times 1\frac{1}{4}$

Exercise 8: Evaluate $2\frac{1}{10} \times \frac{5}{7}$

Division of fractions is a little more tricky but again there is a simple rule to help you. Division by a fraction is exactly the same as multiplication by the 'upside down' fraction. You might see this better by considering $1 \div \frac{1}{2}$. We are really saying here 'How many halves are there in 1?' The answer is 2 and using the rule we get:

$$1 \div \frac{1}{2} = 1 \times \frac{2}{1} = 2$$

Similarly what is $4\frac{1}{2} \div 1\frac{1}{2}$?

$$4\frac{1}{2} \div 1\frac{1}{2} = \frac{9}{2} \div \frac{3}{2} = \frac{9}{2} \times \frac{2}{3} = \frac{18}{6} = \frac{9}{3} = 3$$

which you will see is correct if you ask yourself 'How many $1\frac{1}{2}$s are there in $4\frac{1}{2}$?.

Example: Evaluate $6 \div \frac{9}{10}$

$$6 \div \frac{9}{10} = \frac{6}{1} \div \frac{9}{10} = \frac{6}{1} \times \frac{10}{9} = \frac{60}{9}$$

Exercise 9: Evaluate $\frac{2}{3} \div \frac{4}{3}$

Exercise 10: Evaluate $\frac{4}{3} \div \frac{2}{3}$

Exercise 11: Evaluate $1\frac{1}{2} \div 2\frac{1}{4}$

To end with, let me combine several of these processes in one example:

Exercise 12: Evaluate $\dfrac{(1\frac{1}{2} + 2\frac{1}{4})}{1\frac{1}{2} \times 1\frac{1}{2}} - \frac{2}{3}$

BODMAS

A general rule for evaluating complicated expressions such as exercise 12 is given by a rule known as BODMAS. This represents the order in which the problem is evaluated.

(1) Brackets
   Of
(2) Division

(3) Multiplication

(4) Addition

(5) Subtraction

Therefore, in exercise 12, the bracket $(1\frac{1}{2} + 2\frac{1}{4})$ is first evaluated. Next the denominator term of $1\frac{1}{2} \times 1\frac{1}{2}$ is calculated. The result of the evaluation of the bracket is then divided by the resultant multiplication. Finally this last result has the $\frac{2}{3}$ subtracted to give the final answer.

3

Negative Numbers

In the introduction I referred to negative numbers and to
-100 feet as a height, to $-6\ ^\circ C$ as a temperature, and to
$-£10$ as a (temporary?) bank balance. This is the usual way
of writing such measurements and the numbers included in
them, but it is a source of confusion. Can you see why?

When we speak about subtraction we use a 'minus' sign
$(-)$, so to illustrate the action of taking 6 from 9 we write
$9 - 6$. This sign is placed *between the two numbers* which we
happen to be using, and it is an instruction to *do something*
specific to those two numbers. So $9 + 6$ means that we are to
add 6 to 9; 9×6 refers to multiplication; $9 \div 6$ implies
division, and $9 - 6$ is the way of saying 'from 9 take 6 away'.
Now suppose we go on using the same minus sign as in
$-6\ ^\circ C$. The sign obviously cannot mean the same as before
because previously it meant 'subtract' *from another number*
and if we say simply -6 then there *is not* another number!

In fact there is another number implied but we normally
do not write it; it is the number 0. So -6 is really a shortened
way of writing $0 - 6$. Thus a sub-zero temperature of $-6\ ^\circ C$
is really a temperature of $(0\ ^\circ C - 6\ ^\circ C)$.

The numbers -6 and -10 are '*negative numbers*' and it is
useful to picture them (as on a thermometer) in the following
way.

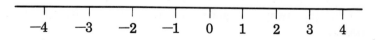

For added emphasis I shall refer henceforth to the 'ordinary' numbers as the *positive numbers*, and indicate them by 6, 10 and so on. Thus —6 and 6 may both refer to £6, but the former represents a sorry state and the latter a happy one! The number zero sits aloof in the middle of the scale and is counted as neither + nor —.

You may be asking why I am being so fussy about this. As we found with fractions, we wish to be able to *use* negative numbers, to *calculate* with them, to get answers to problems which involve negative (as well as positive) numbers. To do this we must know how they behave in arithmetic.

Let us suppose you were in debt for the amount £3. We could say that you had —£3. So if I then gave you £4, how much would you then have? I think you will agree that you would have £1. So —3 + 4 = 1. You could think about this on the chart above, where the numbers refer to intervals on the line.

But what about the case when your bank balance showed you had —£5, that is you were £5 in the 'red'. If you signed a cheque for £10, how much would you then have? You would be further into the red by £10, that is you would have —£15 (assuming the cheque did not bounce and you had a kind bank manager). So

$$(-5) - (10) = (-15)$$

I have put brackets around the 'amounts' to make it clearer.

Now what about subtraction which involves debt? For instance, I have in my pocket a bill for £16 which I have to pay. When I offer to pay it I am told I have been overcharged by £2. What is my position? Initially I owed £16, which I represent as —16. £2 of this debt should not be there, so I want to remove it. From a debt of £16, represented by —16 I want to remove, or subtract, a debt of £2, presented by —2,

$$\text{i.e. } (-16) - (-2)$$

which leaves me with a debt of only £14. This idea of 'removing a debt leaves you richer' can be expressed as the general rule 'taking away a negative is equivalent to adding on its positive'.

Example: What is the value of −20 subtracted from −35?

$$\text{i.e. } (-35) - (-20) = -35 + 20$$

$$= -15$$

Now try the following exercises

Exercise 13: $(-2) + 5 = ?$

Exercise 14: $(-5) + 2 = ?$

Exercise 15: $-2 - (-3) = ?$

In general you see that I avoid a 'bare' succession of two signs. In Exercise 15 we do not have $-2 - -3$; we prefer to keep the -3 self-contained as (-3). On the other hand, when the question starts with a negative (as in Exercise 13 or 14) we please ourselves whether we supply a bracket or not, and usually we do not.

Exercise 16: $-5 + (-2) = ?$

Exercise 17: $-12 + 12 = ?$

Exercise 18: $9 - 8 = ?$

Exercise 19: $8 - 9 = ?$

Exercise 20: $6 - 15 = ?$

Exercise 21: $6 - (-7) = ?$

Exercise 22: $-6 - (-7) = ?$

Having dealt with addition and subtraction we must now look at multiplication and division. These processes are easily summed up by the following rules:

$$(+) \times (+) = (+)$$
$$(+) \times (-) = (-)$$
$$(-) \times (+) = (-)$$
$$(-) \times (-) = (+)$$

Also if a minus sign is outside a bracket and you remove the brackets, then the signs throughout the bracket are changed.

Example: $(-2) \times (-3) = +6$

and $(-2) \times (+3) = -6$

Try the following to get used to the rules of multiplication

Exercise 23: $-2 \times 7 = ?$

Exercise 24: $2 \times 7 = ?$

Exercise 25: $2 \times (-7) = ?$

Exercise 26: $-12 \times 3 = ?$

Exercise 27: $-9 \times 7 = ?$

Exercise 28: $9 \times (-7) = ?$

Exercise 29: $(-2) \times (-7) = ?$

Exercise 30: $-3 \times (-4) = ?$

Exercise 31: $-3 \times (-3) = ?$

Exercise 32: $2 \times (-8) = ?$

Exercise 33: $-8 \times (-2) = ?$

Exercise 34: $17 \times (-1) = ?$

Exercise 35: $-1 \times (-1) = ?$

For division you only need to use the rules of multiplication backwards. We saw that $(-2) \times (+3) = (-6)$, so we can deduce that $(-6)/(-2) = +3$, that is, the rule is that a negative number divided by a negative number is a positive number. In a similar manner $-6/(+3) = -2$.

Exercise 36: $-12 \div (-2) = ?$

Exercise 37: $-12 \div 2 = ?$

Exercise 38: $12 \div (-2) = ?$

Exercise 39: $4 \div \dfrac{4}{3} = ?$

Exercise 40: $-4 \div (-\dfrac{4}{3}) = ?$

There is just one other place where the rules for signs apply, and we really should look at it. Consider the problem of evaluating $22 - (9 - 7)$. The brackets are there for a purpose, it is not the same question as evaluating $22 - 9 - 7$.

The first of these expressions says that from 22 you take $(9 - 7)$, that is 2, so the answer is 20. The second expression says that from 22 you take 9, leaving 13, and then you go on to subtract 7 more, leaving 6. So, if there are *no* brackets, then you proceed from left to right: if there *are* brackets and you take the 9 away first (leaving 13) you will agree that you have taken away 7 too many, since as was shown previously you should only have taken away 2. To put things right you must restore that 7, so in fact $22 - (9 - 7)$ can be evaluated as $22 - 9 + 7$. You see the effect of the minus sign which was in front of the bracket? It effectively changes the minus sign which was in front of the 7 to a plus sign. 'Two minuses making a plus' again? Yes, in a way.

Try another one of the same type. Suppose we have to evaluate $12 - (3 + 4)$. This means that we are to take 7 altogether from 12, leaving 5. If we try to evaluate without brackets we would start with $12 - 3$ which is 9; then what about the 4? We have to *subtract* 4 in order to arrive at the correct answer of 5. So in fact we can see that $12 - (3 + 4)$ is actually equivalent to $12 - 3 - 4$. What about the effect of the minus sign? It has in effect changed the +4 into —4. So, when we write $12 - (3 + 4)$ we mean that *both* 3 *and* 4 are to be taken from 12, so it is not too surprising that an alternative form for it is $12 - 3 - 4$. Are the following questions correct?

Exercise 41: $20 - (6 - 1) = 20 - 6 + 1$

Exercise 42: $20 - (6 + 1) = 20 - 6 - 1$

Exercise 43: $6 - (20 - 19) = 6 - 20 + 19$

Exercise 44: $8 - (7 - 9) = 10$

Exercise 45: $7 - (2 - 3) - 4 = 4$

Exercise 46: $(7 - 2) - (3 - 4) = 6$

Exercise 47: $7 - (2 - 3 - 4) = 12$

Exercise 48: $(7 - 2 - 3) - 4 = -2$

You may be wondering why we bother to express $20 - (6 - 1)$ in any other way. After all we can always do it as $20 - 5$ which is 15, without trying to rewrite as $20 - 6 + 1$. The real point will become apparent in due course when we find it necessary to deal with expressions of a similar type but without a specific knowledge of the numbers. For instance, at a later stage (when we introduce algebra) we shall have to deal with expressions like $12 - (x + 3)$ where x is a number which is unspecified at that stage. It will be useful then to call upon the rule we have just looked at to rewrite $12 - (x + 3)$ as $12 - x - 3$ and hence as $9 - x$. Likewise $3 - (2 - x)$ can be expressed as $3 - 2 + x$ or $1 + x$ even if we do not know the particular value of x. For the moment we have done enough.

Exercise 49: There is just one number which can be *multiplied* by any number at all (positive, negative, whole or fractional) without affecting the second number. What is the number?

Exercise 50: There is just one number which can be *added* to any number at all (positive, negative, whole or fractional) without affecting the second number. What is the number?

SUMMARY

By now I hope you feel familiar with the arithmetic of fractions and negative numbers, because this has been the main purpose so far. There has also been a less obvious but more mathematical purpose: to bring all sorts of numbers into the system, all of them having the same behaviour under addition, multiplication, etc. Because they all behave in the same way, we can now start using letters to represent numbers when it suits us to do so: and this idea will take us a very long way indeed.

LETTERS TO REPRESENT NUMBERS

Look first at the following statements:

(i) $5 = 5 + 3$
(ii) $6 = 2 + 3$
(iii) $5 = 2 + x$
(iv) $y = 2 + x$
(v) $y = w + x$

Each of these statements is an equation in the sense that we are using the 'equals' sign to connect two expressions. The first equation simply says that '5 is equal to 2 plus 3' which we can easily accept. The second says that '6 is equal to 2 plus 3' which is a false statement. We shall try to concern ourselves with statements which are true. So what about equation (iii)?

In equation (iii) we see a letter x which presumably represents a number which is not specified at this stage. Can the letter x represent the number 3? Can x represent the number 4? The answer is 'YES' in each case. If x represents 3 then equation (iii) is actually stating that $5 = 2 + 3$ which is true. If x represents 4 then equation (iii) is actually stating that $5 = 2 + 4$ which is false. Could x represent 5? Yes, but the resulting statement would then be $5 = 2 + 5$ which is false. The point I am making is that although the letter x can represent *any* number at all, there is only *one* such number which makes the equation into a *true* statement, and it is fairly clear that we are more interested in *that* value for x than in all the other possibilities. When we talk about 'solving an equation', we are always interested in every value of x which 'fits' the equation, that is, which makes it into a true assertion.

Exercise 51:   What values must x represent in order to make

$$\text{(i)} \quad 3 + x = 7$$
$$\text{(ii)} \quad x + 7 = 1$$
$$\text{(iii)} \quad x - 2 = 1$$

into true assertions? That is, *solve* the three equations.

Now look at equation (iv) of our original list, that is, $y = 2 + x$. This time we have two letters x and y, each of which represents a number. Could x represent *any* number? Yes. Could y represent any number? Yes. But many of the choices for x and y will give a false statement, e.g. if we chose x to be 5 and y to be 3, then equation (iv) would assert that $3 = 2 + 5$ which is false.

There are values of x and y in these equations which act as 'variables' in the sense that we can 'vary' the values we attach to the letters. Furthermore among the variety of values we *could* choose, there are quite a number of possible choices which produce true statements, and it is these 'suitable' values of x and y which we are usually interested in.

Exercise 52: Look again at $y = 2 + x$. What values of y would be 'suitable' if we chose x to be:

  (i)   9
  (ii) —5
  (iii) 1¾

and of x if, instead we choose y to be:

  (iv) 11
  (v)   —1
  (vi) —2½

You can see from this last example that the equation $y = 2 + x$ really 'fixes' the *connection between* values of x and values of y. Once x has been chosen then the suitable value of y is actually determined. If you specify x as 21 then you have no *choice at all* for a suitable value for y, it *must* be 23 if a *true* assertion is to be made. Equally, if you fix y as 0 then x has to be —2, otherwise the equation is not 'satisfied'.

Exercise 53: Take $y = 2 + x$ again.

  (i)   If x is 12, What value must y have
        if the equation is to be 'satisfied'?
  (ii) If x is —12 what must y be?
  (iii) If y is —12 what must x be?

This last discussion has made a very important point, that equations are a way of describing a connection or relation between two (or more) quantities (represented here by x and y). We sometimes say that $y = 2 + x$ is a 'formula' for y 'in terms of x'. By this we mean that, if x is fixed at some value, then y is thereby fixed also; in other words, that the y value depends upon the x value. Equally, in this case, the x value depends upon the y value. Each equation or formula connects the 'suitable' values of x and y.

Finally, look at equation (v) of the original list, that is, $y = w + x$. Here we have three letters x, y, w, each of which may represent a number. What choices do we have? Clearly a large number of choices. Suppose I specify that x is to be 3;

what can you say about y? Not very much! What about w?
Again you cannot say much. But you can say that the equation (or formula) does connect the possible values which y
and w can take, once x is fixed. With x as 3 you can assert
that $y = w + 3$ which is a formula for y 'in terms of w'. Of
course, if you were to fix the values of x and w then the
suitable value of y is then determined. With x as 3 and w as
12, y must be 15 if we are to have a true statement. And
equally well if we fixed y as 19 and w as 2, then the third
'variable', w is thereby fixed as 17. The free choice of any
two of the three variables is possible, but the third variable is
then fixed.

Exercise 54:   Take $y = w + x$.

     (i) Could $x = 3, y = 7; w = 5$ make the
         equation true?
    (ii) Could they take values 5, 9, 4?
   (iii) If $x = 1$ and $w = 5$ what value of y
         would be suitable?
   (iv) If $x = 1$ and $y = 7$ what is a suitable
         value for w?

    I have so far looked at a number of equations which may
be true or false depending on what values we choose for the
letters which appear in them. In future I shall concern myself
only with the values of x, y, w etc. which produce *true*
assertions. When I am finding the 'suitable' values which 'fit'
the equations I shall call it 'solving' the equations. Equally,
when I speak of a formula which gives a 'connection' or
'relation' between various letters then I shall wish to find
values which 'fit' the formula. Bear these points in mind in
the following exercises.

Exercise 55:   If $y = x + 6$
           (i) find y when $x = 3, 0, -5$
           (ii) find x when $y = 0, 7, -1$

Exercise 56:   If $x + y = 12$ find y when $x = 2, 3, 4, \ldots, 12$
           successively. (The dots are used to indicate the
           other whole numbers between 4 and 12.)

Exercise 57: Does the formula $y = 12 - x$ describe the same connection as in Exercise 56? (This is called a rearrangement or a transposition of the formula in Exercise 56; we shall discuss it more fully later on.)

You will have noticed probably that the only arithmetical operations so far used have been addition and subtraction. Now we are to look in a little more detail at equations and formulae involving multiplication and division.

Consider first the formula $y = 2x + 3$; what does it mean? The first thing to note is that we write $2x$ as a shorthand for 2 x x, that is, 2 multiplied by x. You can see why we leave out the multiplication sign too — it can get confused with a letter x if you do not write it carefully. So this formula suggests that we are thinking of a number, which we call x for the time being, and which we have multiplied by 2. At this stage we have $2x$. To this number (still not specified of course!) we add 3, and the final total is then represented by y. Try it with a specific value for x, say $x = 5$. First we multiply by 2 (giving 10), then we add 3 to the total (giving 13). So if $x = 5$ we find that $y = 13$.

Exercise 58: Using the same formula $y = 2x + 3$, find the value of y if $x = 3$ and if $x = 7$, and if $x = 0$.

Exercise 59: If you know that $y = 7x - 2$, what is x in terms of y?

Exercise 60: If $y = 7(x - 2)$ what is x in terms of y?

Exercise 61: Can you see why $y = 7(x - 2)$ and $y = 7x - 14$ represent exactly the same connection between x and y?

Exercise 62: Here are four equations. See if any of them is the same as any other.

(i) $y = \frac{1}{2}(x - 6)$ that is, $\frac{(x - 6)}{2}$

(ii) $y = \frac{x}{2} - 3$

(iii) $2y = x - 6$

(iv) $x = 2y + 6$

You can see from the last exercise that a given relation or formula connecting x and y may be put into several equivalent forms. The particular forms in (i) and (ii) give y in terms of x; (iii) gives an expression without fractions; (iv) gives x in terms of y. The particular form you use depends upon the circumstances, but the important thing is to be able to re-arrange the form to suit your purposes.

Exercise 63:   Solve $2(x-1)+3=9$

Exercise 64:   Solve $3(x+1)+2=5$

Exercise 65:   Solve $\frac{(x+1)}{2}-1=\frac{1}{4}$

Exercise 66:   Solve $2x-7=5x-9$

Exercise 67:   Solve $2x+7=5x-9$

Exercise 68:   Solve $2x+7=5x+9$

Exercise 69:   Solve $2(x-1)+3=x$

Exercise 70:   Solve $2(x+3)+6=4x$.
[Reminder: $2(x-1)$ means $2x-2$.]

Now here are some revision exercises which cover the solving of equations and the transposing of formulae. When solving the equations do get clear in your own mind this business of the *order* of operations.

4

Revision Exercises: Simple Equations and Transposition of Formulae

R.1.   Solve $x + 5x = 12$

R.2.   Solve $x + 12 = 5x$

R.3.   Solve $5x + 12 = x$

R.4.   Solve $5(x + 12) = x$

R.5.   Solve $\frac{(x - 3)}{4} = 1$

R.6.   Solve $2(x + 1) + 1 = x$

R.7.   Solve $2[2(x + 1) + 1] + 1 = 6$

Re-arrange the following expressions in order to make the symbol which follows each example the subject of the formula:

1. $P = \frac{N + 2}{D}$, N

2. $k = \frac{brt}{v - b}$; b

3. $C = 2\pi r$; r

4. $P = aW + b$; W

5. $S = \frac{n}{2}(a + l)$; (i) n
   (ii) l

6. $A = P + \frac{PRT}{100}$; P

7. $S = 2\pi r(r + h)$; h

8. $v^2 = u^2 + 2as$; (i) s
   (ii) u

9. $L = \frac{Wh}{a(W + P)}$; W

10. $\frac{L}{E} = \frac{2a}{R - r}$; R

11. $R = \sqrt{\left(\dfrac{ax - P}{Q - bx}\right)}$; x

12. $D = \sqrt{\left(\dfrac{3h}{2}\right)}$; h

13. $S = 4\pi r^2$; r

14. $T = 2\pi \sqrt{\left(\dfrac{I}{MH}\right)}$; M

15. $A = \frac{1}{2}m(v^2 - u^2)$; u

16. $d = a^3 \sqrt{\left(\dfrac{H}{N}\right)}$; H

17. $M = \dfrac{Wd}{4}(l - \dfrac{d}{2})$; l

18. $S = \dfrac{Wd}{l}(l - \dfrac{d}{2})$; l

19. $H = \dfrac{(T - t)\pi\, Rn}{275}$; t

20. $H = \dfrac{W}{2}(R^2 - r^2)$; r

21. $T = \sqrt{\left(\dfrac{Pbh}{4 + a^2}\right)}$; (i) b
(ii) a

22. $x = \dfrac{a + 2b}{3(a + b)}h$; a

23. $v = w\sqrt{a^2 - x^2}$; x

24. $A = \pi r^2 \sqrt{(h^2 + r^2)}$; h

25. $T = H + \dfrac{W^2 l^2}{4}$; (i) H
(ii) l

26. $\dfrac{1}{u} + \dfrac{1}{v} = \dfrac{2}{f}$; u

27. $V = \frac{1}{3}\sqrt{\left(\dfrac{S^3}{8\pi}\right)}$; S

28. $A = \sqrt{\left(\dfrac{P^2 - 2Q^2}{2P^2 - Q^2}\right)}$; P

29. $A = \pi r\sqrt{(h^2 + r^2 + \pi r^2)}$; h

30. $r = \dfrac{f}{2} + \sqrt{\left(\dfrac{f^2}{4} + q^2\right)}$; q

31. $e = \dfrac{L - l}{LT - lt}$; L

32. $E = \dfrac{2Hg - V^2}{2Hf}$; H

33. $v = \sqrt{\left[gd(1 + \dfrac{3h}{d})\right]}$; d

34. $T = 2\pi\sqrt{\left(\dfrac{h^2 + k^2}{gh}\right)}$; k

35. $C = \dfrac{ab\sqrt{2g}}{\sqrt{(a^2 + b^2)}}$; a

36. $E = \dfrac{m}{2g}(v^2 - u^2)$; u

37. $2c(1 + \dfrac{1}{m}) = 3k(1 - \dfrac{2}{m})$; m

38. $D = \sqrt{\left(\dfrac{2v^2 d}{g} + \dfrac{d}{4} - \dfrac{d}{2}\right)}$; d

39. $e = \sqrt{\left[\dfrac{t - k}{K(1 + kt)}\right]}$; k

40. $T - W = \dfrac{Wv^2}{32x}$; W

5

Quadratic Expressions

The type of equation I have so far dealt with is usually known as a *'linear'* equation because of its connection with a straight line graph. I shall explain this connection later, but here our main purpose is to give understanding of algebraic manipulation and another task awaits us, namely the type of equation known as the *quadratic* equation. To prepare for that, attempt the following questions.

Exercise 71: What is the area of a rectangle whose length is 6 cm and whose width is 4 cm?

Exercise 72: If you found the previous answer by saying that the area is 6 x 4 square cm, that is, 24 sq. cm, can you explain *why* you multiply 6 x 4 in order to get the answer? (Do not say that you did it 'because that is the rule'! *Why* is it the rule?)

Exercise 73: If a rectangle has length x cm and width y cm, what is its area?

Exercise 74: If a square has a side of length x cm, what is its area?

You can see that we are dealing here with a problem which involves multiplication. The answer to Exercise 73 requires us to find the result of multiplying x by y, and Exercise 74 requires x to be multiplied by x. Now we can write 'x multiplied by y' in the form x x y, and we usually leave out the multiplication sign and write this as xy (or sometimes $x.y$ where the dot simply reminds us that we are multiplying but avoids possible confusion with the letter x itself). Similarly in Exercise 74 we can write x x x as the answer but we may put this in the form $x.x$ or xx. It is usual in this case to write x^2, which we pronounce as 'x squared'. You can see that if we start with x and draw a square of side x cm, then the area of the square is sensibly called x^2 (that is x squared).

Exercise 75: What is the area of a rectangle of length $3x$ cm and width $2y$ cm?

Exercise 76: What is the area of a square whose side measures $3x$ cm.

The only problem in these exercises is 'how do we express the answer'? Exercise 75 requires us to multiply $3x$ by $2y$, and we could write this as $3x$ x $2y$ or $3x.2y$ or simply $3x2y$. But what does this mean? In full it means 3 x x x 2 x y and we can re-write this as 3 x 2 x x x y because it makes no difference in which order we carry out the various multiplications. (Try it if you have any doubts. Calculate 3 x 2 x 3 x 5 and 3 x 3 x 2 x 5 and see that the answers are the same.) So the answer can be expressed as 6 x x x y or more simply as $6xy$. Similarly, Exercise 76 requires us to evaluate $3x$ x $3x$ which we prefer to write in the form $9xx$ or $9x^2$.

Exercise 77: Find the area of a rectangle of length $5x$ cm and width $3y$ cm.

Exercise 78: Find the area of a rectangle of length $5x$ and width $(3x + 4)$ cm.

Again the question is, what form does the answer take? In Exercise 78 we know we have to multiply $5x$ by $(3x + 4)$ that is, $5x$ x $(3x + 4)$ or usually $5x(3x + 4)$. If you use the last form you must remember that it *is* a multiplication which is to be done! There is an alternative way of evaluating the answer which you can see from a diagram:

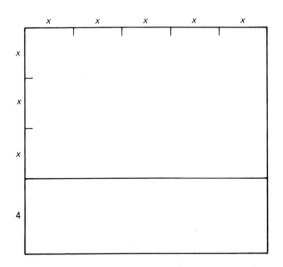

The diagram shows a length of $5x$ and a width of $(3x + 4)$. Of course we do not know what specific value x has at this stage, but the illustration tells us that the total area consists of two rectangles.

Exercise 79:   What are the two separate areas in the figure?

From this, you can see that the total area of $5x(3x + 4)$ is expressible as $5x3x + 5x4$, that is $15x^2 + 20x$. You can also probably see that we do not have to draw a diagram every time, because you can 'see' where the $15x^2$ and the $20x$ come from by looking at $5x(3x + 4)$.

Exercise 80:   Is it true that $2x(3x + 1)$ is the same as $6x^2 + 2x$?

Exercise 81:   Is $(3x + 2)4x = 12x^2 + 8x$?

Exercise 82:   Express $x(x + 2)$ in a different form as the sum of two terms.

Exercise 83:   Likewise express $2x(2 + x)$ in a different form.

6

Expanding Brackets

When we express $5x(3x + 4)$ in the form $15x^2 + 20x$ we say that we have 'expanded' $5x(3x + 4)$, or that we have 'removed the brackets' (that is, removed the *need* for the brackets).

Exercise 84: Expand $x(3x - 1)$

Exercise 85: Expand $3x(1 - x)$

Exercise 86: Draw a rectangle which has a length of $(2x + 1)$ and a width of $(x + 2)$. What is its area? Can you evaluate the area as four separate areas? Do you agree that $(2x + 1)(x + 2) = 2x^2 + 1x + 4x + 2$?

Notice that when brackets are close to each other we intend to indicate a multiplication. Just as xy means x x y, so $(2x + 1)(x + 2)$ means $(2x + 1)$ x $(x + 2)$.

Exercise 87: Could we simplify the last answer to $2x^2 + 5x + 2$?

Exercise 88: Expand $(2x + 1)(3x + 2)$. If you need to draw a diagram then do so — but can you see how to find the four separate 'areas' without actually drawing anything?

You can probably see from these examples that when we have to multiply two brackets together, we get a number of separate terms. In fact each part of one bracket has to be multiplied by each part of the other. So, $(2x + 3)(4x + 1)$ will produce four separate parts:

$$2x \times 4x = 8x^2$$
$$2x \times 1 = 2x$$
$$3 \times 4x = 12x$$
$$3 \times 1 = 3$$

and these four terms can be reduced to three because we can replace $2x + 12x$ by $14x$. So, $(2x + 3)(4x + 1) = 8x^2 + 14x + 3$

Exercise 89: Expand $(x + 5)(x + 6)$

Exercise 90: Expand $(x + 5)(4 + 3x)$

Exercise 91: Expand $(3 + x)(4 + x)$

Exercise 92: Expand $(a + b)(c + d)$

Now that we can 'expand' brackets we can extend the idea to take account of negative terms too. In Exercise 85 you saw that $3x(1 - x) = 3x - 3x^2$, and the only point at which care must be exercised is in connection with the proper signs: are the signs + or —?

Example: $(x - 1)(x + 2)$ will produce four terms. Think of these four terms as being the four parts of the answer to Exercise 92, but note that where we had the letter b in that exercise, we have the number —1 in this example. So instead of $ac + ad + bc + bd$ we shall have:

$$xx + 2x - 1x - 1 \times 2$$

Note that we have had to say 1 x 2 because if we put two *numbers* close together we might think we mean 12 (twelve)! So, $(x - 1)(x + 2) = x^2 + x - 2$ (Can you see where the x term comes from? We had $2x - 1x$ which is the same as $1x$ which is simply x.)

Exercise 93: Expand $(x - 2)(x + 3)$

Exercise 94: Expand $(x - 2)(x - 3)$ [You need to recall the rule which says that $(-2) \times (-3) = +6$]

Exercise 95: Expand $(2x - 1)(x - 3)$

Exercise 96: Expand $(x + 3)(x - 3)$

Exercise 97: Expand $(x + y)(x - y)$

7

Restoring Brackets

Suppose now that we try to reverse the process of 'expanding' brackets. If we see the expression $x^2 - 6x + 5$ is it possible to express this in bracket form as (   )(   ) using whole numbers? Well, sometimes it can be done and sometimes not.

Exercise 98:   Try to express $x^2 - 6x + 5$ in brackets form. (You may have to guess!)

Exercise 99:   What about $x^2 - 5x + 6$?

Exercise 100: What about $x^2 + 5x + 6$?

Exercise 101: What about $x^2 + 2x + 3$?

Exercise 102: What about $x^2 - 4$?

Exercise 103: More difficult would be $12x^2 + x - 6$. Can you see why it is more difficult? (It *can* be done.)

Examples like Exercise 103 are rather tiresome because there are so many apparent possibilities to start with: for example

$$
\begin{array}{ll}
(12x \quad) & (x \quad) \\
(6x \quad) & (2x \quad) \\
(4x \quad) & (3x \quad)
\end{array}
$$

are all perfectly good ways of making sure that you provide the 12 which is necessary. Then you have to make sure that —6 is provided as well; so if you start with $(12x \quad)(x \quad)$ you can then try:

	$(12x \quad 6)$	$(x \quad 1)$
or	$(12x \quad 3)$	$(x \quad 2)$
or	$(12x \quad 2)$	$(x \quad 3)$
or	$(12x \quad 1)$	$(x \quad 6)$

Even then you may find that it is *not* possible for any of these possible forms to ensure that the $+ x$ term is correctly produced. It can be a very tedious business, and why should we bother? Because we are thereby in a position to solve *quadratic equations* (and indeed other more complicated equations also).

Exercise 104:   We can solve $3x + 4 = 2(x - 1)$ easily enough. That is, we can find a value for x which 'fits' the equation. Can you find (by trial and error) a value of x which fits $x^2 + 2x - 3 = 0$?

Exercise 105:   Can you find *another* value of x which *also* fits $x^2 + 2x - 3 = 0$? One of the facts of life about quadratic equations is that often there are *two* different values which x may take, either of which fits the question. We can see how the tedious business of 'putting brackets in' may pay off, by trying the following exercises.

Exercise 106:   Is it true that $x^2 + 2x - 3 = (x - 1)(x + 3)$?

Exercise 107:   If $x^2 + 2x - 3 = 0$ must it also be true that $(x - 1)(x + 3) = 0$?

Exercise 108:   If two numbers like $(x - 1)$ and $(x + 3)$ when multiplied together give an answer of zero, must it be true that at least *one* of the numbers must be zero?

Exercise 109:   So, if $(x - 1)(x + 3) = 0$ is it true that either $(x - 1) = 0$ or $(x + 3) = 0$?

Exercise 110:   If $(x - 1) = 0$ what value must x have? If $(x + 3) = 0$ what value must x have?

Exercise 111:   Now check! If $x = 1$ is true that $x + 2x - 3$ $= 0$? And if $x = -3$? In other words, are these values of x the values which fit the equation $x^2 + 2x - 3 = 0$?

So we can see from this that if we can express a quadratic expression like $x^2 + 2x - 3 = 0$ in bracket form, then we can certainly find two values which x can take and which fit the equation. In other words, we can *solve* the equation.

Exercise 112:   Express $x^2 - 7x + 12$ in bracket form. Solve $x^2 - 7x + 12 = 0$ by repeating the sequence followed in Exercises $106 - 111$.

Exercise 113:   Solve $x^2 - x - 6 = 0$

Exercise 114:   Solve $2x^2 - x - 1 = 0$

Exercise 115:   Solve $6x^2 - 13x - 5 = 0$ [Hint — start with $(3x\quad)(2x\quad)$.]

There will be more about solving quadratic equations in chapter 15. There we shall see how to deal with solving these equations by use of formulae.

SUMMARY

You are now well acquainted with that mixture of numbers, letters repesenting numbers, + and − signs, and brackets, which itself represents a number and which is called an *algebraic expression*. You can change its form without altering the number it represents, for instance by getting rid of brackets. When there is an 'equals' sign separating two expressions you have an *algebraic equation* or *formula*. In many cases you can find which values for the letters in an equation make it true. These values are called *solutions* of the equation, and solving equations is an important aspect of algebra.

Another valuable skill is the manipulation of equations which contain two variables such as x and y. You know how to turn such an equation (if it is linear) into a formula for x, or a formula for y, whichever you like.

8

Graphs and Co-ordinates

We usually use graphs to represent neatly a lot of data. A graph is a special sort of accurate drawing, and points on a graph can be located by what we call *co-ordinates*. Since we will only be working in two dimensions it is usual to locate points on a graph by means of the *x co-ordinate*, that is the distance from the vertical axis, and the *y co-ordinate*, that is the distance from the horizontal axis.

Exercise 116:   Determine the *x* co-ordinate and the *y* co-ordinate of each of the points A, B, C, D shown in figure 1.

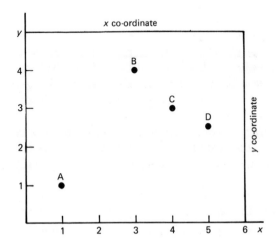

Figure 1.

Now let us try it the other way round. In the next question the co-ordinates of the points are given and you have to mark the position of the points on the diagram given. Marking the position of a point is called *plotting* the point.

Exercise 117:   On the diagram given (figure 2) plot the points A, B, C, and D whose co-ordinates are given as follows:

Point A: (x co-ordinate is 1, y co-ordinate is 2)

Point B: (x co-ordinate is 2, y co-ordinate is 2)

Point C: (x co-ordinate is 4, y co-ordinate is 3½)

Point D: (x co-ordinate is ½, y co-ordinate is 4)

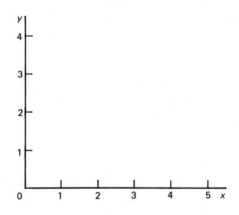

Figure 2

Exercise 118:   Where would a point be if its x co-ordinate were 0?

Exercise 119:   Plot the following points on the diagram given (figure 3)

Point L: (x co-ordinate is 0, y co-ordinate is 3)

Point M: (x co-ordinate is 2, y co-ordinate is 0)

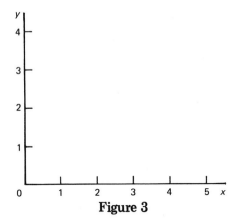
Figure 3

Exercise 120:   What are the co-ordinates of the *origin* — the point we label 0?

NOTATION

You will have no doubt noticed by now that it is rather laborious to write out '*x* co-ordinate is' and '*y* co-ordinate is' each time, so we invent a shorthand notation. Instead of writing: (*x* co-ordinate is 3, *y* co-ordinate is 4), we simply write (3, 4).

So that if we wish to specify a point all we do is to write down a pair of numbers in brackets; the first number is the *x* co-ordinate of the point, the second number the *y* co-ordinate. As you see more written mathematics you will often see 'A is the point (1, 2)' or 'the point (1, 2)'. (Notice the comma to separate the co-ordinates.)

Using this notation we can write down the co-ordinates of the points shown in Exercise 116 as follows:

Point A: co-ordinates (1, 1)

Point B: co-ordinates (3, 4)

Point C: co-ordinates (4, 3)

Point D: co-ordinates (5, 2½)

Notice that here we have written 'co-ordinates (1, 1)'. We could equally well have said 'A is the point (1, 1)' or simply 'A (1, 1)'.

Exercise 121:   Write down the co-ordinates of the points marked in figure 4.

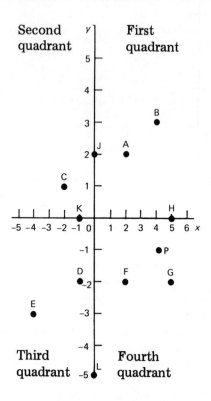

Figure 4

For the next exercise you will need to draw your own axis system on graph paper; this is a straightforward task and if you wish you can simply copy the system we have drawn for Exercise 121, although you do not have to use the scale we have used. You can choose whatever scale you wish. It is usual to use the same scale for both the x and y axes although the finished product would then look rather different to figure 4 above.

Exercise 122: Draw an axis system and mark on the following points: A (4, 2), B (5, 0), C (0, 1), D (−2, 4), E (3, −1), F (−4, −5), G (0, −2), H (−5, 0).

Exercise 123: On your axis system for Exercise 122, put in the following points: K (2, −1.5), L (0.3, −3.3), M (2.3, −4.5), N (−3, 3.8).

Look at the graph below. The points A, B, C and D are all points on a straight line. What has happened to the values of the co-ordinates if you move from point C to point D?

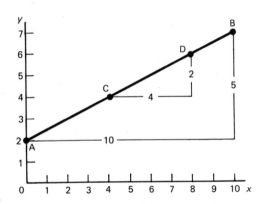

Figure 5

In this case in moving from C to D, y has increased by 2 for an increase of 4 in x. So that for an increase of 1 in x, y increases by 2/4 = 1/2. We sometimes call the change in y as we move from one point to another the 'rise', while the change in x is referred to as the 'run'. We usually combine these into what is known as the *gradient*, where,

$$\text{gradient} = \frac{\text{rise}}{\text{run:}}$$

Example: If we consider points A and B we get:

$$\text{rise} = 5, \text{run} = 10$$

$$\therefore \text{gradient} = \frac{5}{10} = \frac{1}{2}$$

Exercise 124:   Consider points C and B on figure 5. Find the 'rise' and the 'run' and hence the gradient of the line.

Notice that the gradients between points C and D, A and B, and C and B are all equal to ½. Since all the points A, B, C and D lie on a straight line then we can say that the gradient of the straight line equals ½. In fact you could choose any two points on the straight line and the answer would be the same.

Notice in the case of the line AB the 'rise' is always positive since y is increasing as x increases. In the following exercise you will notice that as we move in the direction of x increasing, y decreases, so that the 'rise' is in fact a fall and this is shown as a negative rise.

Exercise 125:   Find the co-ordinates of the points S and T. Find the 'rise' and 'run' and hence the gradient of the line.

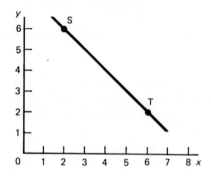

Figure 6

Exercise 126: The diagram given below illustrates a gradient of '1 in 10'. How would we represent such a gradient by a single number?

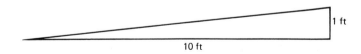

Figure 7

Note

Please do not get what we call here a gradient confused with the gradients you see on road signs, for example 1 in 10. These refer to a rise of 1 foot over a distance along the road of 10 feet, that is

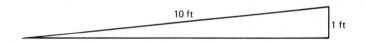

Figure 8

Exercise 127: Find the gradient of the four straight lines shown in figures 9 — 12. (Note: these diagrams are examples of *graphs*.) (Hint: in each case choose two points on the line and find 'rise' and 'run'.)

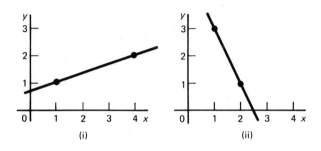

Figure 9   **Figure 10**

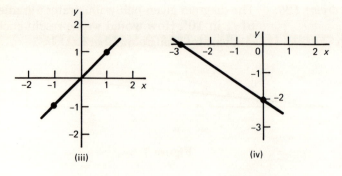

(iii)   (iv)

Figure 11   **Figure 12**

9

Equation of a Straight Line

So far we have seen that a straight line can be described as a line with a constant gradient and we know how to find the gradient of any particular line from a graph of the line. In this section we shall look more closely at the co-ordinates of points on a straight line to see if there is a way of finding each y from each x co-ordinate. If we can establish such a connection then we shall have what is known as the *equation* of the line.

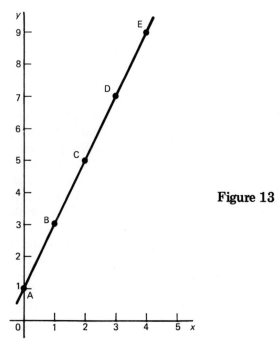

Figure 13

Consider the line shown in figure 13. What is its gradient? .... Answer 2. Now look at the points marked. Let us write down their co-ordinates:

A : (0, 1) B : (1, 3) C : (2, 5) D : (3, 7) E : (4, 9)

You might like to add a few more points of your own.

Can we now find a connection between the numbers in the 2nd column (the y co-ordinates) and those in the first (x co-ordinates)?

The first thing to notice is that the x co-ordinate increases by 1 at each step as we go from A to E, and as we would expect the y co-ordinate increases by 2 each time, since 2 is the gradient of the line (remember gradient is the change in y for an increase of 1 in x). Can you now find the rule which connects y with x? The answer is that to get the y co-ordinate you double the x co-ordinate and add 1. Since we refer to the y co-ordinate simply as 'y' and the x co-ordinate as 'x', the rule is : y is $2 \times x + 1$ or $y = 2x + 1$.

Check a few more points on the line, for example, for fractional values of x, to satisfy yourself that all points have co-ordinates connected by this rule, and since the line extends into the third quadrant (see figure 4) it might be useful to redraw it on a full axis system.

This is all straightforward if the rule is simple to find, but it might not be. There is a more systematic way of finding the equation of a line. Refer back to figure 13 again. The line cuts the y axis at the point A, that is, where y is 1, and having found this point we can use the gradient to find the co-ordinates of any other point. The gradient tells us the change in y for an increase of 1 in x, so in our case since the gradient is 2 the y co-ordinates of points on the line increase at the rate of 2 for each increase of 1 in x (or change in y is twice the change in x).

So if we start at A whose x co-ordinate is zero, moving to B means an increase of 1 in x; hence y increases by 2 so the y co-ordinate of B is $1 + 2 = 3$. (Check this.) Moving from A to C, x increases by 2, so y increases by $2 \times 2 = 4$, so the y co-ordinate is $1 + 4 = 5$. (Check this.) What about D? From A to D the change in x is 3 so the change in y is $3 \times$ gradient $= 3 \times 2 = 6$; hence the y co-ordinate of D is 7.

Notice that by starting (at A) on the y axis the change in x when we move to a point is simply the x co-ordinate of the point; this would not be so if we started at any other point. Hence if we take a point with an x co-ordinate of 10, say, then the change in x from A is 10, so the change in y is 10 x 2 = 20 giving a co-ordinate of 21 for this point.

Hence we have thoroughly tested out rule (for the first quadrant)

$$y = 2x + 1$$

Notice what happens if the point in question has an x co-ordinate which is negative, say −2, then $y = 2$ x (−2) + 1 = −3. (Check that this point, namely (−2, −3) is also on the line.) So we are satisfied that the rule works properly for *all* values of x. We can analyse the rule:

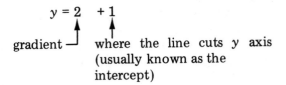

gradient ⌐ where the line cuts y axis
(usually known as the
intercept)

Exercise 128:   Find the equations of the seven lines shown on the graphs (figures 14 − 20).

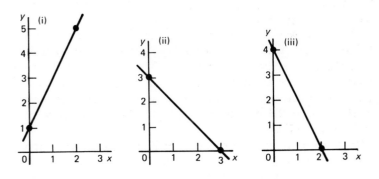

Figure 14   **Figure 15**   **Figure 16**

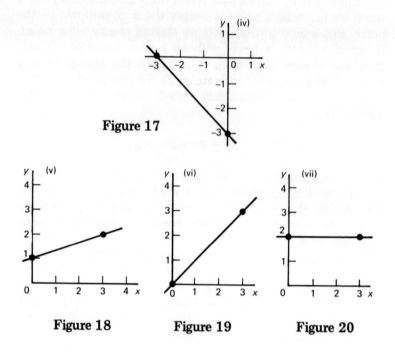

Figure 17

Figure 18   Figure 19   Figure 20

10

Equation to Graph

In chapter 9 we investigated the problem of finding the equation of a straight line given the graph of the line. In practice we often need to reverse this step — to draw a graph of a straight line whose equation is given. From our present knowledge this is a relatively elementary operation. Let us approach the problem via an example: suppose we wish to draw a graph of the line $y = \frac{3}{4}x + 2$.

From our previous work we recognise that the y intercept is 2 and the gradient is $\frac{3}{4}$. So to draw the graph we start at the point $y = 2$ on the y axis and draw a line with a gradient of $\frac{3}{4}$. (From the point $y = 2$ construct a run of 4 and a rise of 3.) The finished graph is shown in figure 21 and is a line with gradient $\frac{3}{4}$, crossing the y axis at the point $y = 2$.

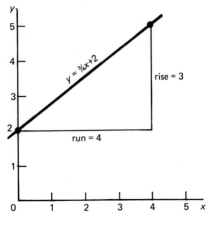

Figure 21

Exercise 129:   On the same axis system draw the graphs of the lines $y = x$, $y = x + 2$, $y = x + 4$, $y = x - 2$, $y = x - 4$.

Exercise 130:   On the same axis system draw the lines: $y = \frac{1}{2}x + 2$, $y = 2x + 2$, $y = 2 - x$.

Exercise 131:   Draw the following straight lines:

(i) $y = 2x - 3$

(ii) $y = 4 - \frac{1}{2}x$

(iii) $3y = x + 3$

(iv) $2y = 5 - 3x$

11

Solving Simultaneous Equations Graphically

Let us summarise this method of solution. Given a pair of simultaneous linear equations, to find a solution we first draw the two straight lines represented by the equations, find the point where they cross and read off the co-ordinates. These co-ordinates then give us our solution. Try the method out on the next exercise — if you do not appear to be getting a solution, ask yourself why?

Exercise 132: Find a solution to the following pairs of simultaneous linear equations:

(i) $3y - 4x = 1,$ $2y + x = 8$

(ii) $4y + 10x = 2,$ $2x - y = 4$

(iii) $2y - 3x = 2,$ $4y + 6x = 4$

(iv) $2y - 3x = 2,$ $4y - 6x = 4$

12

Solving Simultaneous Equations Algebraically

Exercise 133: Below is an incomplete derivation of the solution to the simultaneous linear equations $6y + 4x = 1$, $3y - 8x = 33$. Fill in the missing expressions in the boxes.

The steps in the method involve the transposing and scaling of equations. The numbers above the boxes are to enable you to check your answers. The problem is:

Solve (that is, find x and y that fit) the equations

$$6y + 4x = 1 \qquad (1)$$
$$3y - 8x = 33 \qquad (2)$$

Step 1: Isolate one of the variables, in this case,

$$\text{From (1), } 6y = 1 - 4x \qquad (3)$$

From (2), $3y = \boxed{}^{1}$ $\qquad (4)$

Step 2: Assume a solution, say $x = a$, $y = b$. This means that from equations 3 and 4

$$6b = 1 - 4a \qquad (5)$$

$3b = \boxed{}^{2}$ $\qquad (6)$

Step 3: Make the isolated terms the same in both equations. In this case we can make both '*b*' terms into 6*b* by multiplying equation (6) by 2 giving equation (7) below.

$$6b = 1 - 4a \qquad (5)$$

$$6b = \boxed{}^{\,3} \qquad (7)$$

Step 4: Obtain an equation with only one letter, *a*

$$1 - 4a = \boxed{}^{\,4} \qquad (8)$$

Step 5: Find the value of *a*

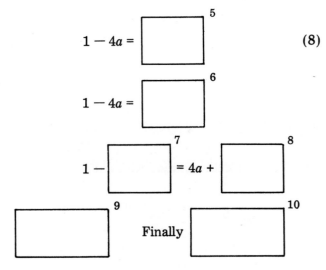

$$1 - 4a = \boxed{}^{\,5} \qquad (8)$$

$$1 - 4a = \boxed{}^{\,6}$$

$$1 - \boxed{}^{\,7} = 4a + \boxed{}^{\,8}$$

$$\boxed{}^{\,9}$$

Finally $\boxed{}^{\,10}$

Step 6:   Find b, from equations (5) and (6) by substituting

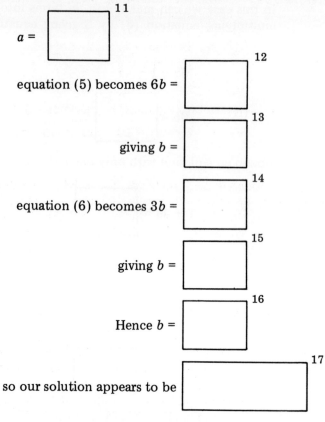

$a =$ ☐ 11

equation (5) becomes $6b =$ ☐ 12

giving $b =$ ☐ 13

equation (6) becomes $3b =$ ☐ 14

giving $b =$ ☐ 15

Hence $b =$ ☐ 16

so our solution appears to be ☐ 17

Step 7:   Check this solution with equations (1) and (2)

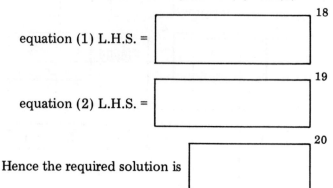

equation (1) L.H.S. = ☐ 18

equation (2) L.H.S. = ☐ 19

Hence the required solution is ☐ 20

Exercise 134: Solve the following pairs of simultaneous linear equations, algebraically:

$$\text{(i)} \qquad 3y - 4x = 1, \qquad 2y + x = 8$$
$$\text{(ii)} \qquad 4y + 10x = 2, \qquad 2x - y = 4$$
$$\text{(iii)} \qquad 2y - 3x = 2, \qquad 4y + 6x = 4$$
$$\text{(iv)} \qquad 4x + 2y = 5, \qquad 3x - 4y = 1$$
$$\text{(v)} \qquad 2x + 7y = 10, \qquad 3x + 5y - 4 = 0$$

Conclusion

I have considered two methods for solving simultaneous linear equations. One method is graphical, the other algebraic. The latter method is usually the one preferred since it will always yield an exact solution if a solution exists. This algebraic method outlined is not the only algebraic method available; there are others which vary slightly, but the basic idea is the same.

RECIPROCALS $\frac{1}{x}$

Exercise 135: How could you show fairly easily that the graph $y = 1/x$ is *not* a straight line (that is, that the function is not linear? (Hint: choose some values of x and calculate the values of y and plot them on a graph.)

Exercise 136: Complete the following tabulations for this 'reciprocal'

x	¼	½	1	2	4
$\frac{1}{x}$	4			½	

x	−¼	−½	−1	−2	−4
$\frac{1}{x}$	−4				

(You can see why we call it the 'reciprocal', if you find the reciprocal of ¼ you get 4 and if you find the reciprocal of 4 you get ¼.)

Exercise 137:   Sketch the graph using your results to Exercise 136.

Exercise 138:   Can the graph which you have sketched be *extended* (You have to ask yourself whether the graph can be extended to right *and* left *and* up *and* down.)

Exercise 139:   What happens when $x = 0$?

Let us now review this rather special curve.

(i)   Obvious points, easily plotted? We did *some* calculations and could have managed with fewer.

(ii)   'Top' or 'bottom' points? There are none in this case, because the graph progresses smoothly (continuously) downwards as x changes from very small (positive) values to larger (positive) values; and it also moves smoothly downwards from left to right for negative values of x. There is no point at which the curve turns round (that is, stops coming up and starts coming down or stops going down and starts coming up).

(iii)   The behaviour as x becomes very large? The further we go to the right (using positive values of x), the lower the curve drops, but it never reaches the horizontal axis. In a similar fashion, as we move left (using negative values of x) so the graph edges nearer to the horizontal axis from below, and likewise never reaches that axis.

(iv)   The behaviour as x becomes very small? The smaller x becomes the larger the value of $1/x$. In fact as x becomes near to 0 we are unable to draw it because the graph becomes too large.

(v)   Note that there is no point on the graph which refers to $x = 0$ because the value of $y = 1/0 =$ infinity.

13

Revision: Equations: Graphs

EQUATIONS

I Basic Arithmetic and 'Rules'

1. Evaluate the following selecting your answer from those given:

    (i) $3 + 7 \times 4$ (a) 40 (b) 31 (c) 84

    (ii) $6 \times 5 - 2 + 4 \times 6$ (a) 52 (b) 42 (c) 18

    (iii) $7 \times 6 - 12 \div 3 + 1$ (a) 40 (b) 39 (c) 21

    (iv) $17 - 2 \times (6 - 4)$ (a) 30 (b) 1 (c) 13

    (v) $3 \times 5 - 12 \div (3 + 1)$ (a) 12 (b) 10 (c) 8

2. Answers to the following questions are either true or false. Indicate the appropriate answer for each problem by evaluating each problem:

    (i) $-5 - 6 = 11$ (iv) $-(-8) \times 5 = 40$

    (ii) $-8 + 3 = -5$ (vii) $-3 \times (-4) = -12$

    (iii) $-7 - (+5) = 12$ (viii) $8 \div -2 = -4$

    (iv) $-5 - (-8) = 3$ (ix) $-9 \div -3 = 3$

    (v) $(-6) \times (-7) = 42$ (x) $(-6)^2 = 36$

II Basic Algebra

3.   *Formulae*

(i)   If a train travels S metres in t seconds find the distance it travels in x minutes.

(ii) If a m^2 of sheet metal has a mass of x kg what will be the mass of b m^2?

(iii) If $a = 1$, $b = 2$, $c = -3$, find the values of

$$3a + b - c \text{ and } \frac{5bc}{a}$$

(iv) If $x = 5$, $y = 2$ and $z = 3$ find the values of

$$\frac{3x - x}{2x - 3y} \text{ and } \frac{3x + 7y}{8z}$$

(v)   Find the values of the following:

$A = \pi r l$ when $\pi = \dfrac{22}{7}$, $r = 4$ and $l = 14$

$I = \dfrac{E}{R}$ when $E = 240$ and $R = 5$

$K = \dfrac{Wv^2}{g}$ when $W = 100$, $v = 25$ and $g = 10$

(vi) The time of swing of a simple pendulum is given by the formula $t = 2\pi \sqrt{\left(\dfrac{l}{g}\right)}$. Find t when $l = 90$ and $g = 10$. Take $\pi = 3.14$

(vii) Find the value of V from the formula $V = E - IR$ when $E = 220$. $I = 15$ and $R = 0.75$

4.   *Algebraic Simplification*

Simplify the following:

(i) $7x + 11x$

(ii) $-2x + 4x$

(iii) $8a - 6a - 7a$

(iv) $6ab - 3ab - 2ab$

(v) $14xy + 5xy - 7xy + 2xy$

(vi) $3x - 2y + 4z - 2x - 3y + 5z$

(vii) $3a \times 4b$

(viii) $x \times (-y)$

(ix) $(3a) \times (-4b) \times (-c) \times 5d$

(x) $-3a \div -3b$

(xi) $4ab \div 2a$

(xii) $(-12a^2 b) \div 6$

(xiii) $7a^2 b^2 \div 3ab$

(xiv) $-3pq \times -3q$

(xv) $7ab \times -3a^2$

(xvi) $3(3x + 2y)$

(xvii) $-(a + b)$

(xviii) $-3y(3x + 4)$

(xix) $-2m(-1 + 3m - 2n)$

(xx) $5(2a + 4) - 3(4a + 2)$

(xxi) $3(x - 4) - (2x + 5)$

(xxii) $5(2x - y) - 3(x - 2y)$

(xxiii) $4(1 - 2x) - 3(3x - 4)$

(xxiv) $3(a - b) - 2(2a - 3b) + 4(a - 3b)$

(xxv) $3x(x^2 + 7x - 1) - 2x(2x^2 + 3) - 3(x^2 - 5)$

5.   *Brackets*

Expand the following brackets and simplify:

(i) $(x + 1)(x + 2)$

(ii) $(2x + 5)(x + 3)$

(iii) $(5x + 1)(2x + 3)$

(iv) $(x - 1)(x - 3)$

(v) $(2x - 1)(x - 4)$

(vi) $(x - 8)(4x - 1)$

(vii) $(3x - 1)(2x - 5)$

(viii) $(x + 3)(x - 1)$

(ix) $(2x + 5)(x - 2)$

(x) $(3x + 5)(2x - 3)$

(xi) $(3x - 5)(2x + 3)$

(xii) $(2p - q)(2p + q)$

(xiii) $(5a - 7)(5a + 7)$

(xiv) $(3x + 4y)(2x - 3y)$

(xv) $(2x + 3)^2$

(xvi) $(x - 1)^2$

(xvii) $(2a + 3b)^2$

(xviii) $(a - b)^2$

(xix) $(3x - 4y)^2$

(xx) $x(x - y)^2$

6.   *Miscellaneous Problems*

In each of the following questions one answer is either 'true' or 'false'. Indicate the appropriate word for each question:

(i)   The value of $8ab \div 3c$ when $a = 6$, $b = 4$, $c = 2$ is 32

(ii) $-3 \times -7 \times -8 \times -2$ is equal to -336

(iii) $8x - 5x$ is equal to 3

(iv) $15xy + 7xy - 3xy - 2xy$ is equal to $17xy$

(v) $5x^3y^2z$ is the same as $5y^2zx^3$

(vi) $4x(3x - 2xy)$ is equal to $12x^2 - 8x^2y$

(vii) $-8a(a - 3b)$ is equal to $8a^2 + 24ab$

(viii) $3(x - y) -5(2x - 3y)$ is equal to $12y - 7x$

(ix) $2x(x - 2) -3x(x^2 - 5)$ is equal to $-3x^3 + 2x^2 -19x$

(x) $3a(2a^2 + 3a - 1) -2a(3a^2 + 3)$ is equal to $9a^2 + 3a$

III Simple Equations

7. In the following questions the answer is either 'true' or 'false'. Indicate the appropriate word for each problem:

    (i)   If $\frac{x}{7} = 3$ then $x = 21$

    (ii)   If $x - 5 = 10$ then $x = 5$

    (iii)   If $x + 8 = 16$ then $x = 2$

    (iv)   If $x - 7 = 14$ then $x = 21$

    (v)   If $3x + 5 = 2x + 10$ then $x = 3$

    (vi)   If $2(3x + 5) = 18$ then $x = 1$

    (vii)   If $2(x + 4) - 5(x - 7) = 7$ then $x = 12$

    (viii) If $6y = 10(8 - y)$ then $y = 15$

    (ix)   If $\frac{x}{2} - 1 = \frac{x}{3} - \frac{1}{2}$ then $x = 3$

    (x)   If $\frac{3}{x + 5} = \frac{4}{x + 2}$ then $x = 26$

    (xi)   If $\frac{3}{x - 6} = \frac{2}{x - 2}$ then $x = -2$

    (xii)   If $\frac{2x - 3}{2} = \frac{x - 6}{5}$ then $x = 3$

8. In the following questions indicate the letter(s) corresponding to the correct answer(s).

    (i)   If $3(2x - 5) - 2(x - 3) = 3$ then x is equal to (a) 3 (b) 6 (c) ¾ (d) 1¼

    (ii)   If $\frac{x - 5}{3} = \frac{x + 2}{2}$ then x is equal to (a) 16 (b) −16 (c) 7 (d) −7

    (iii)   If $3(x - 2) - 5(x - 7) = 12$ then x is equal to (a) −8½ (b) 8½ (c) −7 (d) 0

    (iv)   If $\frac{3 - 2y}{4} = \frac{2y}{6}$ then y is equal to (a) $\frac{18}{20}$ (b) $\frac{9}{10}$ (c) 3 (d) −3

IV Simultaneous Equations

9. In the following give the letter(s) corresponding to the correct answer(s):

(i) In the simultaneous equations $2x - 3y = -16$, $5y - 3x = 25$, $x = -5$. Hence the value of y is :

(a) 2 (b) -2 (c) $\frac{26}{3}$ (d) 0

(ii) By eliminating x from the simultaneous equations $3x + 5y = 2$ and $x + 3y = 7$ the equation obtained is (a) $4y = 19$ (b) $8y = 9$ (c) $y = 19$ (d) $4y = 5$

(iii) The solutions to the pair of equations $2x - 5y = 3$ and $x - 3y = 1$ are

(a) $x = -3$, $y = 22$ (b) $x = 4$, $y = 1$
(c) $x = 3$, $y = 2$ (d) $x = 3$, $y = -2$

(iv) Two numbers, x and y, are such that their sum is 18 and their difference is 12. The equations which will allow x and y to be found are:

(a) $x + y = 18$, $y - x = 12$
(b) $x + y = 18$, $x - y = 12$
(c) $x - y = 18$, $x + y = 12$
(d) $y - x = 18$, $y + x = 12$

(v) A motorist travels x km at 50 km/h and y km at 60 km/h. The time taken is 5 hours. If his average speed is 56 km/h then:

(a) $50x + 60y = 5$ (b) $6x + 5y = 1500$

$\qquad x + y = 280$ $\qquad x + y = 280$

(c) $\dfrac{x}{50} + \dfrac{y}{60} = 5$ (d) $50x + 60y = 5$

$\qquad x + y = 56$ $\qquad x + y = 280$

GRAPHS, INCLUDING SOLUTIONS OF EQUATIONS

1. The extension, in centimetres, produced by a load, in newtons, for a certain spring is given in the following table:

Load (N)	2	4	6	8	10	12
Extension (cm)	1.4	2.7	4.3	5.5	7.0	8.3

Represent this data graphically and draw the straight line which most nearly fits the points you have plotted. From the graph find the load required to give an extension of 5.0 cm. Use the graph to estimate the extension of the spring produced by loads of (i) 20 N (ii) 100 N
Are these estimates valid?

2. Draw the graph of:

$$y = 3x - 2$$

for values of x in the range

$$-1 \leqslant x \leqslant 4$$

From the graph, identify:

(i)   the gradient of the line
(ii)   the point of intersection with the x axis (that is, when $y = 0$)
(iii)   the point of intersection with the y axis (that is, when $x = 0$)

3. A slow train starts from London for Carlisle at 10 o'clock and travels at a constant speed of 60 km/h. One hour later,

a faster train leaves London for Carlisle at 96 km/h. Find graphically the distance from London at which the second train overtakes the first.

4. A man starts out from Bolton at 09.00 h walking to Manchester at 6 km/h. After one hour he meets the bus which runs from Manchester to Bolton. The bus waits a quarter of an hour at Bolton and then returns to Manchester. The bus always travels at 24 km/h. Find when and where the man is overtaken by the returning bus.

5. Draw the graph of:

$$P = 3r + 5$$

for the values of r between 0 and 3.
Use this graph to solve the equation:

$$3r + 5 = 0$$

6. Use a graph for the range $0 \leqslant x \leqslant 6$ to solve the equation

$$3(x - 2) = 6$$

7. Draw the graph of:

$$y = x^2 + 9x + 20$$

for values of x between -8 and 2. Hence solve the quadratic equation:

$$x^2 + 9x + 20 = 0$$

8. Draw the graph of:

$$y = x^2 - 8x + 32$$

for $0 \leqslant x \leqslant 8$. Use the graph to find the values of x for which $y = 18$. Determine the solutions of the equation:

$$x^2 - 8x + 14 = 0$$

9   Use a graphical method to solve the simultaneous linear equations:

$$3x + 2y = 12$$

and

$$4x - 3y = -1$$

10. Draw a graph for the two simultaneous equations:

$$y = x^2 - x + 7$$

and

$$y = x + 8$$

Identify the points of intersection and determine the quadratic equation which has these values as its roots.

11. Draw a graph to represent:

$$y = x^2$$

and

$$y = x + 1,$$

and hence solve the equation $x^2 - x - 1 = 0$

14

Quadratic Equations

These are equations which involve terms in x^2 (that is x x x).
An example of such an equation is $y = x^2$

Exercise 140: Evaluate $y = x^2$ for $x = 1$, $x = -1$, $x = 3$

Exercise 141: Complete the following table of values for the equation $y = x^2$

x	-3	-2	-1	0	1	2	3
y		+4			1		

Sketch the graph using these values.

Exercise 142: Can the graph be extended beyond this information? What happens to y as x grows larger towards the right. And what happens for large *negative* values of x?

The shape we have just drawn — the basic quadratic graph — is called a *parabola*, and in fact all quadratic graphs have the same basic shape.

Exercise 143: On the same diagram show the graphs of $y = x^2$,

$$y = 2x^2, \; y = \tfrac{1}{2}x^2$$

Even though the graphs may be flatter or steeper than the original, the shapes are the 'same'. In fact the shape can be modified by a change of scale on one or other of the axes, but we still refer to the quadratic graph as a parabola whether it takes a steep or shallow form.

What effect will the inclusion of extra terms have? For instance what does $y = x^2 + 2$ look like? Clearly we will be adding 2 on to all the values which we evaluated in Exercise 141, and the effect will be that the whole graph is moved up the page by 2 units. Now let us see the effect of including terms which contain x.

Exercise 144: Consider the equation $y_1 = (x - 1)^2$ and compare it with $y_2 = x^2$.
Evaluate these at $x = 2$ for y_1 and at $x = 1$ for y_2.
If you sketch y_1 on the same diagram as y_2 how would the two graphs be positioned relative to each other?

For this last exercise we can see that — just as $y = x^2$ and $y = x^2 + 2$ are the 'same' curve but the second is 2 units *higher* than the first — so $y = x^2$ and $y = (x - 1)^2$ are the 'same' curve, but the second is one unit to the *right* of the first. (Be careful here — you may have expected it to be to the left!)

Exercise 145: Compare the graphs of the three equations $y = x^2, y = (x + 1)^2, y = x^2 + 1$ by sketching them on the same diagram.

Exercise 146: Consider the equation $y = x^2 + 2x$. Do you agree that an equivalent form for this equation is $y = (x + 1)^2 - 1$?

Exercise 147: Sketch $y = x^2$. On the same diagram sketch $y = (x + 1)^2$. [Refer back to Exercise 145 to remind yourself of the effect of having an $(x + 1)$ instead of an x^2, and to the text immediately after Exercise 143 for the effect of adding (or subtracting) a number.]

Exercise 148: Show that $x^2 + 4x - 1$ is equivalent to $(x + 2)^2 - 5$. Use the same ideas as in Exercise 146 to determine the effect of having $(x + 2)^2$ instead of x^2, and of having a 5 subtracted. Sketch the curve which you end up with when these effects are accounted for.

Exercise 149: Sketch the graph of the equation $y = x^2 - 2x + 3$, by rewriting in the form $(x - ?)^2 + ??$

By this time you will be agreeing with the comment made earlier to the effect that all these quadratic graphs have the same shape (apart from an element of stretching or movement vertically or horizontally). You can also see that any quadratic equation can be rearranged (as in Exercise 149) so as to indicate what vertical or horizontal movement is needed. Sometimes we can use an extra device, illustrated in Exercise 150.

Exercise 150: Show that $y = x^2 - 2x - 3$ may be rewritten as $y = (x + 1)(x - 3)$. Evaluate y at $x = 0$, $x = -1$ and $x = 3$. Use these pieces of information, together with your knowledge of the usual shape of a quadratic graph, to sketch the graph of $y = x^2 - 2x - 3$. Why did we ask in particular for $x = -1$ and $x = 3$? And why $x = 0$?

Exercise 151: Repeat Exercise 150 using the equation $y = 2x^2 + 5x - 3$ in the form $y = (2x - ?)(x + ??)$ (Which particular values of x are most likely to be useful ones here, instead of -1 and 3 which we used in Exercise 150?)

The only extra problem not yet discussed refers to the possibility of an equation like $y = 1 - x^2$ or like $y = -2x^2$ in which the x occurs in conjunction with a negative sign. Let us start with $y = -x^2$; what is the effect on the *graph* of having $-x^2$ rather than x^2?

If you refer back to Exercise 141 you will find tabulated values of x (which was called y at that time). The values of 9, 4, 1, 0, 1, 4, 9 are now to have a negative sign put in front of them if we are evaluating $-x^2$. Can you see the effect this will have?

Exercise 152: Sketch $y = x^2$ and $y = -x^2$ on the same diagram.

Exercise 153: Compare $y = -x^2$ with $y = 2 - x^2$. What is the effect of adding 2?

Exercise 154: Sketch $y = -x^2 + 3x - 2$ by rewriting it in the form $y = (2 - x)(x - ??)$ Is it quadratic? Which way up is the graph — is it a valley or a hill? Any *special* points?

We have done enough on quadratics now. There is basically only one quadratic 'shape' — and it is the shape of $y = x^2$. It may be steeper or shallower; it may be moved to a new position; it may be turned upside down, but it remains a parabola.

15

Solving Quadratic Equations by Formulae

If you can memorise a certain formula you can write down the roots of a quadratic equation with hardly any effort of calculation. Although it looks hard, this formula is one of the two most memorable ones in mathematics. Since it has to apply to *any* quadratic equation we must represent the numerical parts of the three terms by letters, using a, b, and c:

$$ax^2 + bx + c = 0 \qquad (1)$$

You will agree that any quadratic equation can be manipulated into the form of equation (1), with a single number in place of each a, b and c. Any of these numbers may of course be negative. The formula you must remember is:

The roots of equation (1) are $x = \dfrac{-b \overset{+}{-} \sqrt{(b^2 - 4ac)}}{2a}$ $\qquad (2)$

Example: Solve $3x^2 + 2x - 2 = 0$

Looking at this equation in relation to equation (1) we see that a is 3, b is 2 and c is -2. It is a good idea to start by working out $(b^2 - 4ac)$ and this is $(4 - [4 \times 3 \times -2])=28$. So

$$x = \frac{-2 \overset{+}{-} \sqrt{28}}{6}$$

Dividing top and bottom by 2,

$$x = \frac{-1 \pm \sqrt{7}}{3} \text{ or, if you prefer, } x = -\frac{1}{3} + \sqrt{\left(\frac{7}{3}\right)}$$

and $x = -\frac{1}{3} - \sqrt{\left(\frac{7}{3}\right)}$ are the two solutions.

The formula (2) can be proved, simply but messily, by first converting

$$ax^2 + bx + c = 0 \text{ to } x^2 + \frac{b}{a}x = -\frac{c}{a}$$

Next you complete the square by adding $\frac{b^2}{4a^2}$ [which is $\left(\frac{b}{2a}\right)^2$]

to both sides. You obtain $(x + \frac{b}{2a})^2$ on the left-hand side and

$\frac{b^2}{4a^2} - \frac{c}{a}$ which simplifies to $\frac{b^2 - 4ac}{4a^2}$, on the right.

You can probably see where the various terms of formula (2) will come from, so we will leave you to finish the proof if you are so inclined.

Note: (i) the square root of x^2 is both $+ x$ and $- x$, written $\overset{+}{-} x$, [Check $(-x)^2 = ?$] and $\overset{+}{-} x$ are the real roots of $\sqrt{(x^2)}$ — they exist.

(ii) the square root of $-- x^2$, or any negative number, is unreal or imaginary because we cannot take the square root of -1.

(iii) the difference between $(-x)^2$ and $-x^2$.

The three parts of Exercise 155 are repeats, chosen to present you with all aspects of formula (2) in practice. Exercise 156 is to show you that the expression $(b^2 - 4ac)$ is also useful in its own right (hint)!

Exercise 155: Solve by formula

(a) $2x^2 + 3x - 2 = 0$

(b) $2x^2 - 8x + 8 = 0$

(c) $2x^2 + 3x + 2 = 0$

Exercise 156:   Merely prove that

(a) $3x^2 - 2x + \dfrac{1}{3} = 0$ has only one root

(b) $5x^2 - 17x + 15 = 0$ has no real root

Note: *If* a root contains the square root of a negative number, for example, -4 then we will not be able to solve it with the knowledge we have at present. So all we can say is that the root is imaginary.

Exercise 157:   Solve where possible:

(a) $0 = 10 + x^2 - 6$

(b) $2x^2 + 3x - 2 = 0$

(c) $8x = 2x^2 + 8$

(d) $2x^2 + 3x + 2 = 0$

(e) $x^2 - 7x + 11 = 0$

(f) $x^2 + 7x - 11 = 0$

16

Linear Interpolation

We now move on to a different topic, that of Interpolation.

Exercise 158: Suppose you are given two numbers which are the first and last numbers in a sequence of numbers. Suppose these numbers are 6 and 18. If you knew that there were two other numbers *equally spaced* between 6 and 18, could you find the missing numbers? (We take 'equally spaced' to mean that the difference between any two adjacent numbers is the same.)

Exercise 159: Repeat Exercise 158 but on the assumption that there are *three* missing numbers, equally spaced between 6 and 18.

What you have just done is called *Linear Interpolation*. The word *Interpolation* refers to the process of 'filling-in' the missing numbers; the fact that we were trying to provide equal spacing ensured that we were carrying out *Linear Interpolation*.

Exercise 160: If the population of a village was 2000 in 1800 and had dropped to 200 by 1850, what would the population have been in 1825 using *linear* interpolation? Is the use of linear interpolation reasonable?

The point about Exercise 160 is twofold. First of all, we have absolutely *no* reason to assume a linear sequence;

secondly, it is very much *easier* to make that assumption! Linear interpolation is therefore the simplest kind to carry out, but may (in some cases) be wide of the mark. For instance the population of the village in Exercise 160 may have been constant until 1820 and at that time a natural disaster may have suddenly reduced the population.

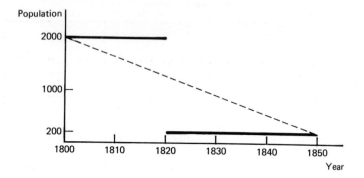

Figure 22

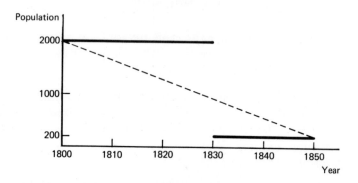

Figure 23

Or the natural disaster may not have taken place until 1830 as in figure 23. So in these two different instances the 'right' answers would be 200 (in the first case) and 2000 (in the second case). The linear interpolation process (indicated by the dotted straight line) gives an answer of 1100 (the *middle* value in a steadily declining value from 2000 to 200).

Having issued this warning, we proceed to use linear interpolation in a number of examples!

The following table gives a set of results of an experiment to measure heat loss. A large beaker of water was heated to boiling point and allowed to cool naturally. The temperature of the water was measured, as accurately as possible, every ten minutes.

Time from start (min)	temp °C	time	temp °C	time	temp °C	time	temp °C
0	100	130	44.73	260	23.40	390	15.17
10	93.64	140	42.28	270	22.46	400	14.81
20	87.74	150	40.00	280	21.58	410	14.47
30	82.25	160	37.88	290	20.76	420	14.15
40	77.14	170	35.91	300	20.00	430	13.86
50	72.40	180	34.08	310	19.29	440	13.59
60	68.00	190	32.38	320	18.64	450	13.33
70	63.90	200	30.80	330	18.03	460	13.10
80	60.08	210	29.33	340	17.46	470	12.88
90	56.56	220	27.97	350	16.93	480	12.68
100	53.27	230	26.70	360	16.44	490	12.49
110	50.21	240	25.52	370	15.99	500	12.31
120	47.37	250	24.42	380	15.57		

Exercise 161:   After 60 minutes the temperature had dropped to 68.00 °C, and after 80 minutes it was
60.08 °C. Estimate the temperature after 70
minutes by linear interpolation.

Exercise 162:   Estimate the temperature after 2 hours by
linear interpolation between the 1 hour and
the 3 hour readings.

Exercise 163:   Is there a lesson to be learnt from the previous
two exercises? What is it?

Exercise 164:   Note the readings after 50 minutes and 60
minutes, and use linear interpolation to
estimate the temperature after (i) 55 minutes,
(ii) 52 minutes.

There is of course scope for the use of common sense in all
this! The cooling curve which can be drawn using the tabulated data is of the following type:

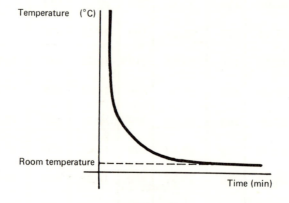

Figure 24

The graph is steepest at the beginning of the experiment
because the heat is more rapidly dissipated then. When the
water is nearly down to room temperature there is little
excess heat to lose.

If linear interpolation is applied between points A and B on such a graph

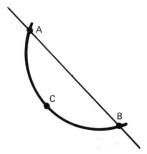

Figure 25

then it is clear from the diagram that the result is likely to be too high as an estimate of the temperature at C. On the other hand, if a curve bends the other way

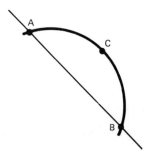

Figure 26

then linear interpolation produces too low an estimate.

In some experiments it may not be at all clear whether the curve is bent one way rather than the other, in which case linear interpolation produces an answer which may be useful but is even more dubious.

Exercise 165:   The following table is a section of a table of natural logarithms:

x	1	1.1	1.2	1.3	1.4	1.5	1.6
ln x	0.000	0.0953	0.1823	0.2624	0.3365	0.4055	0.4700

x	1.7	1.8	1.9	2.0
ln x	0.5306	0.5878	0.6419	0.6931

Using linear interpolation estimate:

(i) ln 1.35 (ii) ln 1.12 (iii) ln 1.83

Note: It is not necessary to have any knowledge of logarithms in order to do this question!

Exercise 166:   Use linear interpolation in connection with the following table of values, to estimate y for $x = 0.7$, $x = 0.65$ and $x = 1.43$

x	y
0.00	0.5000
0.20	0.4207
0.40	0.3446
0.60	0.2743
0.80	0.2119
1.00	0.1587
1.20	0.1151
1.40	0.0808
1.60	0.0548
1.80	0.0359
2.00	0.0228

17

Indices

In an expression like $3^2 = 9$, the ^{2} is called an *index* and 3 is called the *base*. We will now see how to manipulate combinations of numbers which have different indices and the same base. By the end of the section we shall be using any real number as an index, so here is how to read them. Apart from 'squared' for the index ^{2} and 'cubed' for the index ^{3}, the easiest way to read an expression containing an index is the straightforward one. Thus 2^4 and 2^5 are 'two to the four' and 'two to the five'. We sometimes say 'two to the power five' instead. Sometimes the index will be just a letter or an expression containing letters. For example, 10^p is 'ten to the p' and 10^{p-q} is 'ten to the $p-q$'.

You will remember that positive integer indices are used to represent successive self-multiplication. For instance, $3 \times 3 = 3^2$. It would be equally true that $(-3) \times (-3) = (-3)^2$ but to avoid trouble we shall in this discussion not permit bases to be negative, but will allow any positive number as base. Here now is a crucial question in which the base is first any positive number a, and then 10.

Exercise 167: What are $a^4 \times a^3$ and $10^2 \times 10^3$?

The important thing to notice in this exercise is that when *multiplying* numbers which have indices and the same base we just *add* the indices. That is, $4 + 3 = 7$ so the first result is a^7, and $2 + 3 = 5$ so the other result is 10^5. For want of a better name we shall call it the *add : multiply* rule, but remember, the base must be the same.

Exercise 168: Can you simplify $a^4 \times 10^2$?

SQUARE ROOTS : FRACTIONAL INDICES

If the add : multiply rule is to apply to fractional indices then we can see what a fractional index must mean. For example, by the add : multiply rule $3^{\frac{1}{2}}$ x $3^{\frac{1}{2}} - 3^1 = 3$, so $3^{\frac{1}{2}}$ must mean that (positive) number which, multiplied by itself, gives 3. But this is the square root of 3. Thus $3^{\frac{1}{2}} = \sqrt{3}$.

Similarly $5^{\frac{1}{3}}$ x $5^{\frac{1}{3}}$ x $5^{\frac{1}{3}} = 5$, so $5^{\frac{1}{3}}$ is the cube root of 5, that is, the (positive) number x such that $x^3 = 5$. We could write $5^{\frac{1}{3}} = \sqrt[3]{5}$, using the raised ^{3} before the root sign to indicate that the root is the *cube* root.

What about something like $6^{\frac{2}{3}}$? We know that $6^{\frac{1}{3}}$ is the cube root of 6 and also that $6^{\frac{1}{3}}$ x $6^{\frac{1}{3}} = 6^{\frac{2}{3}}$ so that $6^{\frac{2}{3}}$ is the square of the cube root of 6 (or the cube root of the square of 6). Similarly, if a is positive and b and c are positive integers then $a^{b/c}$ means the cth root of a, raised to the power b, that is, $(a^{1/c})^b$, or $[^c\sqrt{(a)}]^b$ or $\sqrt[c]{(a)^b}$.

Generally speaking, most people prefer the index form to the use of root signs. Not always: the roots of $ax^2 + bx + c = 0$ are almost always written as

$$\frac{-b \overset{+}{-} \sqrt{(b^2 - 4ac)}}{2a}$$

rather than

$$\frac{-b \overset{+}{-} [b - 4ac]^{\frac{1}{2}}}{2a}$$

Also, fractional indices may, when convenient, be written as decimal fractions; thus $7^{1\frac{1}{5}}$ can be written as $7^{2.2}$.

Exercise 169: What positive number, multiplied by itself, equals 2^{10}, 2^6, 2?

Exercise 170: Simplify $a^{\frac{1}{3}}$ x $a^{\frac{1}{4}}$; a^2 x $a^{\frac{1}{4}}$; a^2 x a^2 x a^2; $(a^2)^3$.

The only new result to notice is that in evaluating $(a^2)^3$ you do not *add* the ^{2} and the ^{3}: you *multiply* them. This is a consequence of the add : multiply rule. It is a rule in itself, however, and might be called the multiply : power rule.

NEGATIVE NUMBERS AS INDICES

We know that the base is positive. Can we have a *negative* index? If we can, and if we want the add : multiply rule to hold (as we do — it turns out to be so convenient as to be indispensable!) then $a^m \times a^n = a^{m+n}$ no matter whether m or n are positive or negative. Thus, for instance

$$a^5 \times a^{-3} = a^{5+(-3)} = a^{5-3} = a^2$$

and

$$a^{-5} \times a^3 = a^{-5+3} = a^{-2}$$

To make sense of these last two lines, what must a^{-3} and a^{-5} and a^{-2} mean? You will see that they make sense if:

$$a^{-3} = \frac{1}{a^3}, \; a^{-5} = \frac{1}{a^5}, \; a^{-2} = \frac{1}{a^2}$$

for we obtain $a^5 \times a^{-3} = a^5 \times \dfrac{1}{a^3} = \dfrac{a^5}{a^3} = \dfrac{\not{a} \times \not{a} \times \not{a} \times a \times a}{\not{a} \times \not{a} \times \not{a}} = a^2$

and

$$a^{-5} \times a^3 = \frac{1}{a^5} \times a^3 = \frac{a^3}{a^5} = \frac{\not{a} \times \not{a} \times \not{a}}{\not{a} \times \not{a} \times \not{a} \times a \times a} = \frac{1}{a^2} = a^{-2}$$

We take these as examples of a general result: if a is any positive real number and x is any real number:

$$a^{-x} = \frac{1}{a^x}$$

Exercise 171: Suppose in the formula $a^{m-n} = a^m \times a^{-n}$ that a is 10, m is 4 and n is 3. Find the value of a^{-n} ; is it a positive or a negative number?

Notice that, provided the base is positive, then that base with an index is always positive, whatever the index.

Exercise 172:   Zero is a number which cannot be used as a divisor. Is it any use as an index? If so, what can we deduce about a^0, 10^0 etc?

We would have, no matter what integer n may be, $a^0 = a^n \times a^{-n} = a^n \times \dfrac{1}{a^n} = \dfrac{a^n}{a^n} = 1$. If you deduced this for yourself, that is, that any positive number to the power zero equals 1, you did very well.

SUMMARY

The facts about indices which you need to remember are the following. (They are all consequences of the first, but each should be remembered in its own right.)

Laws of Indices

(i) $a^x \times a^y = a^{x+y}$

(ii) $a^x \div a^y = a^{x-y}$

(iii) $(a^x)^y = a^{xy}$

(iv) $\sqrt[y]{(a^x)} = a^{x/y}$

(v) $\dfrac{1}{a^{-1/x}} = \sqrt[x]{a}$

(vi) $a^{-x} = \dfrac{1}{a^x}$

(vii) $a^0 = 1$

Exercise 173:   Evaluate $2^{2.5} \times 2^{3.5} \times 2^{-4}$

Exercise 174:   Evaluate $\dfrac{7^{3.5}}{7^{4.6} \times 7^{-1.1}}$

Exercise 175:   Evaluate $(27)^{4/3}$ and $(16)^{3/2}$

Exercise 176:   Evaluate $16^{3/2} \times \sqrt[3]{(27^2)}$

Exercise 177:   Use what we called the multiply : power rule to evaluate $(10^3)^{2/3}$; $(10^{2/3})^3$; $(a^{m+n})^2$

SCIENTIFIC NOTATION

Very small numbers and very large numbers are not easy to write or to comprehend immediately. An alternative way of writing them is called for — scientific notation. It helps to find logarithms of numbers and also to explain the numbers that appear on scientific calculations. It involves powers of 10. Numbers such as 1000 are easy to write as 10^3, but 2000? Since 2000 is 2 x 1000 then it can be written as 2×10^3.

$$\text{Similarly } 2736 \qquad = 2.736 \times 10^3$$

$$\text{and } 149\ 700 \qquad = 1.497 \times 10^5$$

In this way every number can be written as a decimal which lies between 1 and 10, multiplied by 10 raised to an appropriate power. An easy way to find the power of 10 is to move the decimal point in the number to the left until the number is between 1 and 10, and count the number of steps taken. This is then the power of 10.

Example:

(i)   431.32

   move the decimal point 2 places to the left gives
   4.3132, the power of 10 is 2. So
   $431.32 = 4.3132 \times 10^2$

(ii)   2736

   the decimal point is understood to be to the right of
   the 6, that is, 2736.
   move the decimal point 3 places to the left gives
   2.736
   hence $2736 = 2.736 \times 10^3$

Suppose we have a number smaller than 1, for example, 0.0032. Then we have to move the decimal point to the *right* until we get a number between 1 and 10. This time we write the power as a negative number.

Example:

0.00321

move the decimal place to the right 3 places
then $0.00321 = 3.21 \times 10^{-3}$
and $0.0729 = 7.29 \times 10^{-2}$

Exercise 178: Put the following in scientific notation

- (a)   23065
- (b)   981.665
- (c)   22414
- (d)   273.16
- (e)   0.04129
- (f)   0.1336
- (g)   0.000751
- (h)   0.00000100

Exercise 179: Now put these into decimals

- (a) 2.312×10^3
- (b) 7.95×10^1
- (c) 4.59×10^4
- (d) 6.79×10^6
- (e) 2.29×10^{-8}
- (f) 3.0077×10^{-5}
- (g) 7.46×10^{-1}
- (h) 4.03×10^0
- (i) 3.52×10^{-2}

18

Revision: Indices

1. Write down the value of each of the following:

   10^2, 10^5, 2×10^2, 8^3, 3×8^3, 2.6×10^7, 3.95×10^{10},

   3.95×10^{-10}, 5^0, $(-10)^2$

2. A number, N, can be written in the form $N = a.10^b$. Express each of the following numbers in this form assuming that a must be greater than (or equal to) 1, but less than 10 (that is, $1 \leqslant a \leqslant 10$)

   256   72.95   1,010,110   125,265.629

   0.00109   1.25   1,000,000,000,000

3. Considering the answers you have from 1 and 2 above, write down the advantages of using indices as a shorthand notation for very large or very small numbers.

4. Evaluate the following, expressing your answers in a form using an index:

   $10^2 \times 10^8$ $(2 \times 10^2) \times (5 \times 10^4)$ $(2 \times 10^2) + (5 \times 15)$

   $8^3 \times 8^4$ $2^{10} \times 2^{11}$

   From the above write down how indices are processed when the numbers are multiplied.

5. Evaluate the following:

$$10^2 \div 10^3 \qquad 10^2 \div 10^{-3} \qquad 10^5 \div (2 \times 10^4)$$

$$10^5 \div (2 \times 10^{-9}) \qquad 8^4 \div 2^3 \qquad 8^4 \div 2^{-4}$$

6. Without using indices for your answer, write down the value of the following numbers:

$$100^{1/2} \qquad 100^{-1/2} \qquad 1000^{1/3} \qquad 1000^{-2/3}$$

$$8^{1/3} \qquad 8^{-1/3} \qquad (1/4)^{1/2} \qquad (1/8)^{-1/3} \qquad 16^{1/4}$$

7. Evaluate:

$$(2^3)^5 \qquad (10^2)^3 \qquad (10^2)^{1/2} \qquad (3^4)^{3/4}$$

[Hint: try to determine the rule for evaluating $(a^b)^x$]

19

Logarithms

Part of a printed table of common logarithms and anti-logarithms is shown on p. 82. Most tables look much the same, although in some cases the 12th and 22nd columns (both headed 10) are omitted.

To find the logarithm of a number:

(i)   convert the number into scientific form,
(ii) look up the decimal part of it in the logarithm tables. For example,

187 is the same as 1.87×10^2.

The logarithm of 1.87 is .2718.

(iii) The whole number part of the logarithm is the power of 10 in the scientific form of the original number. For example,

the logarithm of 1.87×10^2 is 2.2718.

Again the logarithm of 1964, the stages are:

(i)   1964 is 1.964×10^3.
(ii) The logarithm of 1.964 is .2931,
(iii) The whole number in front of the logarithm is 3, giving 3.2931.

If the power of 10 is *negative* then the minus sign is placed above the power of 10, for example, to find the logarithm of 4.91×10^{-4} yields $\overline{4}.6911$.

LOGARITHMS

	0	1	2	3	4	5	6	7	8	9	1 2 3	4 5 6	7 8 9
10	0000	0043	0086	0128	0170	0212	0253	0294	0334	0374	5 9 13 / 4 8 12	17 21 26 / 16 20 24	30 34 38 / 28 32 36
11	0414	0453	0492	0531	0569	0607	0645	0682	0719	0755	4 8 12 / 4 7 11	16 20 23 / 15 18 22	27 31 35 / 26 29 33
12	0792	0828	0864	0899	0934	0969	1004	1038	1072	1106	3 7 11 / 3 7 10	14 18 21 / 14 17 20	25 28 32 / 24 27 31
13	1139	1173	1206	1239	1271	1303	1335	1367	1399	1430	3 6 10 / 3 7 10	13 16 19 / 13 16 19	23 26 29 / 22 25 29
14	1461	1492	1523	1553	1584	1614	1644	1673	1703	1732	3 6 9 / 3 6 9	12 15 19 / 12 14 17	22 25 28 / 20 23 26
15	1761	1790	1818	1847	1875	1903	1931	1959	1987	2014	3 6 9 / 3 6 8	11 14 17 / 11 14 17	20 23 26 / 19 22 25
16	2041	2068	2095	2122	2148	2175	2201	2227	2253	2279	3 6 8 / 3 5 8	11 14 16 / 10 13 16	19 22 24 / 18 21 23
17	2304	2330	2355	2380	2405	2430	2455	2480	2504	2529	3 5 8 / 3 5 8	10 13 15 / 10 12 15	18 20 23 / 17 20 22
18	2553	2577	2601	2625	2648	2672	2695	2718	2742	2765	2 5 7 / 2 4 7	9 12 14 / 9 11 14	17 19 21 / 16 18 21
19	2788	2810	2833	2856	2878	2900	2923	2945	2967	2989	2 4 7 / 2 4 6	9 11 13 / 8 11 13	16 18 20 / 15 17 19
20	3010	3032	3054	3075	3096	3118	3139	3160	3181	3201	2 4 6	8 11 13	15 17 19
21	3222	3243	3263	3284	3304	3324	3345	3365	3385	3404	2 4 6	8 10 12	14 16 18
22	3424	3444	3464	3483	3502	3522	3541	3560	3579	3598	2 4 6	8 10 12	14 15 17
23	3617	3636	3655	3674	3692	3711	3729	3747	3766	3784	2 4 6	7 9 11	13 15 17
24	3802	3820	3838	3856	3874	3892	3909	3927	3945	3962	2 4 5	7 9 11	12 14 16
25	3979	3997	4014	4031	4048	4065	4082	4099	4116	4133	2 3 5	7 9 10	12 14 15

ANTILOGARITHMS

	0	1	2	3	4	5	6	7	8	9	1 2 3	4 5 6	7 8 9
·00	1000	1002	1005	1007	1009	1012	1014	1016	1019	1021	0 0 1	1 1 1	2 2 2
·01	1023	1026	1028	1030	1033	1035	1038	1040	1042	1045	0 0 1	1 1 1	2 2 2
·02	1047	1050	1052	1054	1057	1059	1062	1064	1067	1069	0 0 1	1 1 1	2 2 2
·03	1072	1074	1076	1079	1081	1084	1086	1089	1091	1094	0 0 1	1 1 1	2 2 2
·04	1096	1099	1102	1104	1107	1109	1112	1114	1117	1119	0 1 1	1 1 2	2 2 2
·05	1122	1125	1127	1130	1132	1135	1138	1140	1143	1146	0 1 1	1 1 2	2 2 2
·06	1148	1151	1153	1156	1159	1161	1164	1167	1169	1172	0 1 1	1 1 2	2 2 2
·07	1175	1178	1180	1183	1186	1189	1191	1194	1197	1199	0 1 1	1 1 2	2 2 2
·08	1202	1205	1208	1211	1213	1216	1219	1222	1225	1227	0 1 1	1 1 2	2 2 3
·09	1230	1233	1236	1239	1242	1245	1247	1250	1253	1256	0 1 1	1 1 2	2 2 3
·10	1259	1262	1265	1268	1271	1274	1276	1279	1282	1285	0 1 1	1 1 2	2 2 3
·11	1288	1291	1294	1297	1300	1303	1306	1309	1312	1315	0 1 1	1 2 2	2 2 3
·12	1318	1321	1324	1327	1330	1334	1337	1340	1343	1346	0 1 1	1 2 2	2 2 3
·13	1349	1352	1355	1358	1361	1365	1368	1371	1374	1377	0 1 1	1 2 2	2 3 3
·14	1380	1384	1387	1390	1393	1396	1400	1403	1406	1409	0 1 1	1 2 2	2 3 3
·15	1413	1416	1419	1422	1426	1429	1432	1435	1439	1442	0 1 1	1 2 2	2 3 3
·16	1445	1449	1452	1455	1459	1462	1466	1469	1472	1476	0 1 1	1 2 2	2 3 3
·17	1479	1483	1486	1489	1493	1496	1500	1503	1507	1510	0 1 1	1 2 2	2 3 3
·18	1514	1517	1521	1524	1528	1531	1535	1538	1542	1545	0 1 1	1 2 2	2 3 3
·19	1549	1552	1556	1560	1563	1567	1570	1574	1578	1581	0 1 1	1 2 2	3 3 3
·20	1585	1589	1592	1596	1600	1603	1607	1611	1614	1618	0 1 1	1 2 2	3 3 3
·21	1622	1626	1629	1633	1637	1641	1644	1648	1652	1656	0 1 1	2 2 2	3 3 3

From these tables we read off the *decimal part* of a log; thus a number with digits 187 has a log whose decimal part is .2718 (see logarithm table row 18, column headed 7). A number with digits 1876 has a log whose decimal part is .2732. (Obtained by looking up 187 as before, noting down .2718, and then going along the same row to the column headed 6 on the far right, reading off 14 and adding this 14 to the last digit of .2718. These nine or ten columns on the far right, which give the 'corrections' to be added for each fourth digit, are sometimes called the 'mean differences'.) The *whole number* part of the log, as we have seen, does not involve tables: it merely locates the number itself between two successive powers of 10. The smaller index is the required whole number. (Remember, for example, that −2 is smaller than −1.)

Check that you agree with the following examples. Use 'four-figure' tables; a difference of 1 in the last digits between your result and ours may be ignored. (The makers of tables do not always use exactly the same conventions for approximating, and the four-figure common logarithm of any number other than an integral power of ten must be an approximation.)

$$\log 50 \cong 1.6990$$

$$\log 40000 \cong 4.6021$$

$$\log (4 \times 10^6) \cong 6.6021$$

$$\log (4 \times 10^{-3}) \cong \bar{3}.6021$$

$$\log 0.004 \cong \bar{3}.6021$$

$$\log 64 \cong 1.8062$$

$$\log 6.4 \cong 0.8062$$

$$\log 496 \cong 2.6955$$

$$\log 3.27 \cong 0.5145$$

$$\log 0.684 \cong \bar{1}.8351$$

$$\log 0.0000777 \cong \bar{5}.8904$$

$$\log 5.177 \qquad \cong \ 0.7141$$

$$\log 103.3 \qquad \cong \ 2.0141$$

$$\log 107.3 \qquad \cong \ 2.0306$$

$$\log 534697 \qquad \cong \ 5.7199$$
(take the number as 534700)

$$\log 0.005745123 \ \cong \ \bar{3}.7593$$

$$\log 1 \qquad \qquad = \ 0$$

Use log tables 'backwards' and antilog tables (see p. 82) 'forwards' to check that you agree with the following:

The number whose logarithm is 0.8062 is about 6.4
The number whose logarithm is $\bar{1}$.1004 is about 0.126
The number whose logarithm is 4.9084 is about 80990
The number whose logarithm is $\bar{3}$.0836 is about 0.001212

Note that the decimal part of the logarithm is *always positive* and the whole number part only indicates the power of 10.

REVISION

Use log tables for *all* calculations. Do *not* use a slide-rule or electronic calculator.

Section A

1.   Multiply 23.4 by 17.23

2.   Divide 107.2 by 31.9

3.   Find the value of $\dfrac{374.2 \times 2721}{19.3 \times 44.2}$

4.   Find the value of 2.84^3

5.   Find the value of $(2.84^3)^4$

6.   Find the value of $2.84^{3.51}$

7.   Divide 1.0 by 4.0

8.   Multiply 0.3742 by 0.2721

9.   Multiply 0.00193 by 0.0442

10.   Find the value of $\dfrac{0.3742 \times 0.2721}{0.00193 \times 0.0442}$

11.   Divide 23.42 by 4625

12.   Divide 0.0142 by 0.00278

13.   Find the square root of 27.24

14.   Find the value of $\sqrt[4]{0.3726}$

Answers

Q1. 403.1	Q2. 3.360	Q3. 1193.0
Q4. 22.91	Q5. 275,200	Q6. 30.97
Q7. 0.25	Q8. 0.1018	Q9. 0.00008531
Q10. 1193.0	Q11. 0.005064	Q12. 5.109
Q13. 5.219	Q14. 0.7812	

Section B

15.   The $\log_{10}$ of 300 is 2.4771. What is $\log_e 300$?
  ($\log_e 10 = 2.303$)

16.   Find (from the table of natural logarithms) $\log_e 5214$.
  Now (without using the tables of common logarithms)
  determine $\log_{10} 5214$.

  ($\log_e 10 = 2.303$)

17.   By 'taking logs', express the following relationships in
  a form which is similar to that of the 'equation to a
  straight line' that is, $y = mx + c$

    (a) $p = 2^q$

    (b) $m = n^2$

    (c) $h = ak^b$ (note: a and k are constants)

    (d) $Q = C_D^{2/3} B\sqrt{(2g)}h^{2/3}$

This is the equation which relates the rate of flow of liquid, Q, over a weir to the height of the liquid in the weir, (that is, Q and h are the two variables).

C_D is the 'coefficient of discharge' of the weir (a constant)

B is the width of the weir

g is gravitational acceleration

Answers

Q15. 5.706   Q16. (8.5592, 3.717)

Q17. (a) $\log p = q \log 2$

       (b) $\log m = 2 \log n$

       (c) $\log h = b \log k + \log a$

       (d) $\log Q = \frac{2}{3} \log h + \log k$

where $k = C_D^{2/3} B \sqrt{(2g)}$

Section C

18.   Evaluate

      (a) 89.08×0.0003771

      (b) $3771 \div 8.987$

      (c) $4.001 \div 778.3$

      (d) $(0.08724 \times 36.85) \div 791.4$

      (e) $(31.17)^3$

(f) $(0.08005)^{1.7}$

(g) $97.43 \div (0.3524 \text{ x } 6.321)^{2.56}$

(h) $(0.002371)^{-1.47}$

(i) $\sqrt{(0.2407)}$

19. Use indices to express the following:

(a) $\log_3 9 = 2$

(b) $\log_x 1 = 0$

(c) $\log_x\left(\dfrac{1}{a^4}\right) = -4$

20. Find the value of x in the following equations:

(a) $\log_2 8 = x$

(b) $\log_3 (\frac{1}{9}) = x$

(c) $\log_5 0.04 = x$

(d) $4^{x/2} = 3^2$

21. (a) Find the logarithm of $32^5 \sqrt{4}$ to base $2\sqrt{2}$

(b) Express the logarithm of $\dfrac{\sqrt{a^3}}{c^5 b^2}$ in terms of $\log a$, $\log b$ and $\log c$

22. Given $\log 3 = 0.4771$, find $\log 2.7^3 \text{ x } (0.81)^{4/5} \div 90^{5/4}$

23. Given $\log 2 = 0.3010$, find $\log_{25} 200$

24. In calculating compound interest the formula used is

$$A = P\left(1 + \frac{N}{100}\right)^n$$

where P is the money lent and A is what it amounts to in n years at an interest rate of $N\%$ per annum.
If $A = £130$, $P = £100$ and $n = 7$ years, find the rate of interest.

25.   The stress, y, experienced in a shaft rotating n times per second is

$$y = \frac{0.026n^2}{1 - 0.012n^2}$$

What is the critical speed — that is the speed for which y becomes infinitely great.

Answers

18.   (a)   0.0336 (b) 420 (c) 0.00514 (d) 0.00406
      (e)   30300 (f) 0.0137 (g) 12.5 (h) 7230
      (i)   0.491

19.   (a) $3^2 = 9$ (b) $x^0 = 1$ (c) $x = \dfrac{1}{a^4}$

20.   (a) $x = 3$ (b) $x = -2$ (c) $x = -2$ (d) $x = 3.17$

21.   (a) 3.6 (b) $\frac{3}{2} \log a - 5 \log c - 2 \log b$

22. $\bar{2}.27781$

23.   1.646

24.   4%

25.   9.128 r.p.s.

EXAMPLES OF CALCULATIONS

We conclude with an example of calculations using logarithms: follow it through very carefully. We have shown far more details, and used far more space, than you will need when you are competent. The example uses barred logarithms because this is where care is needed. It is a good idea to do a rough check without logs.

Calculate 0.09386 x 0.008472 x 42.3

Number	*Logarithm*	
0.09386	$\bar{2}.9725$	
0.008472	$\bar{3}.9280$	Add
42.3	1.6263	

$$\bar{4} + 2.5268$$

$$= -4 + 2 + 0.5268$$

Answer 0.03363 ◄——— $= \bar{2}.5268$

Check   0.09 x 0.008 x 40

$$= 0.00072 \times 40$$

Answer   = 0.0288 :

compares with 0.03 above

Exercise 180: Use logs to perform the following calculations:

(i)   29.63 x 4.967

(ii)   0.0842 x 31.03

(iii)   27.4 ÷ 195.6

(iv)   0.034 ÷ 0.007962

(v) $(0.0349)^3$

(vi) $\sqrt{7744}$

(vii) $\sqrt[3]{(0.07294)}$

(viii) $\dfrac{23.42 \times 464.7}{5.4 \times 10^6}$

(ix) $\dfrac{17}{8.39} + \dfrac{42.7}{3.26}$

Note: It should be noted, although we do not have space to deal with it here, that logarithms arise in other contexts: they are not merely aids to calculation.

20

Errors and Significant Figures

First of all, by *error* we mean the *approximate value* of something *minus* the *true value*. In so far as these values (approximate and true) are numbers, so is the error a number. It is positive if the approximate value is too high, and negative if the approximate value is too low.

If we are asked to find the square root of 2 [usually written as $\sqrt{2}$ we might say $1.4^2 = 1.96$, $1.5^2 = 2.25$ so that $\sqrt{2}$ is a number between 1.4 and 1.5, nearer to 1.4. More arithmetic would show that $\sqrt{2} \cong 1.4142$ but, as the 'approximately equals' sign indicates, still this is not exact. The statement tells us (or should tell us) that $1.41415 < \sqrt{2} < 1.41425$ where $<$ means less than ($>$ means greater than). Read it as '1.41415 is less than $\sqrt{2}$ which is less than 1.41425'. In fact further computation shows that $x \cong 1.4142135$ so that $1.41421345 < x < 1.41421355$.

The process of rounding off a number to a specific number of decimal places is usually quite straightforward, for example, to four decimal places

$$1.47621 \text{ rounds to } 1.4762$$
$$1.47628 \text{ rounds to } 1.4763$$
$$1.476257 \text{ rounds to } 1.4763$$

However, 1.47625 presents a problem since we are not sure which way it should be rounded. Let us use the convention that when the figure to be shed is 5 exactly, cancel the 5 and add 1 to the next significant figure.

1.47625 rounds to 1.4763

1.47635 rounds to 1.4764

1.4736539 rounds to 1.4737

.6131457 rounds to .6131

We can also talk of rounding to four significant figures in which case we are not concerned with the position of the decimal place, but we want to present the answer correct to four digits, the leading one not being a zero. For example, rounding to four significant figures gives:

1.47625 rounds to 1.476

14762.4867 rounds to 14760

(the last digit is not significant in this sense)

0.0012756 rounds to 0.001276

14.675 rounds to 14.68

Rounding a number of course introduces an error, and you will recall that the absolute error is the approximate value minus the exact value. This error will clearly not always be positive. Thus in the last four cases above the (absolute) errors are:

-0.0025, -2.4867, $+0.0000004$ and $+0.005$

Exercise 181: (a)   Round the following numbers to 4 decimal places (π and e are the names given to two numerical constants you may have heard of.)

(i)   3.1415926 ($\cong \pi$)

(ii)   2.71828 ($\cong e$)

(iii)   22.45913 ($\cong \pi^e$)

(iv)   23.14065 ($\cong e^\pi$)

(v)   65.05145   (vi)   6.71355

(vii)   0.0123456   (viii) 176.43285

(b) Round the above numbers to 4 significant figures

21

Angles, Triangles and Trigonometrical Functions

Let us first define some angles.

Acute angle   this is an angle less than 90°

Right angle   an angle of 90°

Obtuse angle   an angle greater than 90° and less than 180°

Reflex angle   an angle greater than 180° and less than 360°

Complementary   a term used to describe angles whose sum is 90°

Supplementary   a term used to describe angles whose sum is 180°

So that in figure 27 below we would say 'angles *a* and *b* are *complementary*'

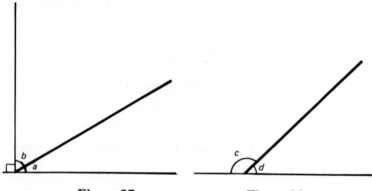

Figure 27   Figure 28

or alternatively '*b* is the *complement* of *a*' (or vice versa). Similarly the other situation in figure 28 could be expressed by stating 'angles *c* and *d* are *supplementary*' or '*c* is the *supplement* of *d*' (or vice versa).

Notice that *a*, *b* and *d* are *acute* angles and *c* is an *obtuse* angle. Note also the 'arc' to mark the angles, a right angle would be marked as ⌐ instead of using an 'arc'.

A *triangle*, or to be more precise, a *plane triangle*, is a figure bounded by three straight lines. (To be correct we need to say a *plane* triangle to distinguish it from a *spherical* triangle, a triangle on the surface of a sphere, which is often used in navigation and has different properties from the plane triangle. We shall omit the word plane from now on.)

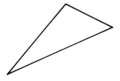

Figure 29

The boundary lines of a triangle are called the *sides* of the triangle. These sides form three angles (hence the name of the figure) and the points where the lines meet, the 'points' of the 3 angles, are called the *vertices* of the triangle. It is usual to label the vertices of a triangle with capital letters as shown below:

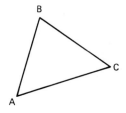

Figure 30

The next figure is an example of an *ISOSCELES* triangle, that is, a triangle with two of its sides of the same length. Such a triangle also has the property that the two angles 'opposite' the equal sides are equal.

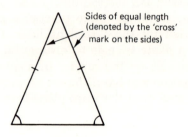

Figure 31

(The name *isosceles* is derived from two Greek words 'isos' meaning *equal* and 'skelos' meaning *leg*; hence literally it means 'equal legs'.) The two equal angles are referred to as *base angles* and the third angle as the *vertical angle*. The side opposite the vertical angle is called the *base*.

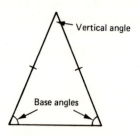

Figure 32

An isosceles triangle has the extra property that if the line which bisects the vertical angle is drawn and extended to cross the base it passes through the *mid-point* of the base and is *perpendicular* to the base. We say that 'the bisector of the vertical angle is the perpendicular bisector of the base'. This angle bisector splits the triangle into 2 equal triangles: the triangles are symmetrical about this line (see figure 33).

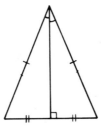

Figure 33

The figure below is an example of an *EQUILATERAL* triangle. This is a triangle which has all of its sides of equal length. It has the property that each of its angles is 60°.

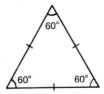

Figure 34

There is another special triangle, one which has one of its angles a right angle so it is called a *right angled* triangle.

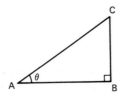

Figure 35

The longest side, the side opposite the right angle, is called the *hypotenuse*. In fact there is a relationship which links together the lengths of the sides of this triangle and it is known as *Pythagoras' Theorem*. It says:

$$AC^2 = AB^2 + CB^2 \text{ for } any \text{ right angled triangle}$$

If we call the angle at A, θ, then we can express what are known as *trigonometric* functions in terms of the sides of the right angled triangle.

$$\text{sine } \theta = \frac{CB}{AC} \text{ usually written as } \sin \theta = \frac{CB}{AC}$$

$$\text{cosine } \theta = \frac{AB}{AC} \text{ usually written as } \cos \theta = \frac{AB}{AC}$$

$$\text{and tangent } \theta = \frac{CB}{AB} \text{ usually written as } \tan \theta = \frac{CB}{AB}$$

The trigonometric functions for all values of θ are listed in standard tables, an extract of which is shown below:

θ°	$\sin \theta$	$\cos \theta$	$\tan \theta$
0	0	1.0000	0
15	0.25882	0.96593	0.26795
30	0.50000	0.86603	0.57735
45	0.70711	0.70711	1.0000
60	0.86603	0.50000	1.73205
75	0.96593	0.25882	3.73205
90	1.00000	0	∞

If you find the values for 0° and 90° strange think about what the appropriate right angled triangle would look like!

Example: In a right angled triangle the hypotenuse is 5 cm long. If one of the angles is 30° what are the lengths of the other two sides?

We have a triangle which looks like:

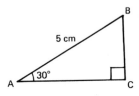

Figure 36

We know that $\sin 30° = \dfrac{BC}{AB} = \dfrac{BC}{5}$

So $\qquad\qquad BC = 5 \times \sin 30°$

We can find the value of $\sin 30°$ from the previous table, its value is 0.5000, hence

$$BC = 5 \times 0.5$$

$$\therefore BC = 2.5 \text{ cm}$$

In a similar manner

$$\cos 30° = \frac{AC}{AB} = \frac{AC}{5}$$

$$\therefore AC = 5 \times \cos 30°$$

$$AC = 5 \times 0.86603$$

$$\therefore AC = 4.33015$$

Exercise 182: If in figure 36 the lengths of the sides were as follows, find the value of the angle at A.

(a) AB = 10 cm,   BC = 5 cm

(b) AB = 10 cm,   AC = 5 cm

(c) BC = 5 cm,   AC = 5 cm

(d) AC = 1.2941 cm,   AB = 5 cm

22

General Revision

1. FRACTIONS

(a) What does $1\frac{7}{8} + 1\frac{1}{4} = ?$

(b) What does $3\frac{1}{4} - 1\frac{7}{8} = ?$

(c) Is it true that $\frac{3}{8} = \frac{9}{24}$?

(d) Write $\frac{10}{12}$ as $\frac{?}{6}$

(e) Evaluate $3\frac{1}{2}$ x $4\frac{1}{2}$

(f) Evaluate $3\frac{1}{8} \div \frac{5}{8}$

(g) Evaluate $\dfrac{(4\frac{1}{8} - 2\frac{1}{4})}{1\frac{1}{2} \text{ x } \frac{3}{4}} - \frac{1}{2}$

2. NEGATIVE NUMBERS

(a) $-3 - (-2) = ?$

(b) $1 - 12 = ?$

(c) $12 - (-1) = ?$

(d) $3 + (12) = ?$

(e) $6 - 7 = ?$

(f) $7 - (-1) = ?$

(g) $-7 - 1 = ?$

(h) $-1 - 7 = ?$

(i) $-2 \times 3 = ?$

(j) $-2 \times (-3) = ?$

(k) $-3 \times 2 = ?$

(l) $3 \div 2 = ?$

(m) $-3 \div 2 = ?$

(n) $3 \div (-2) = ?$

(o) $-3 \div (-2) = ?$

(p) Does $3 - (2 + 1) = 3 - 2 - 1$?

(q) Does $3 - (2 - 1) = 3 - 2 - 1$?

3. LETTERS TO REPRESENT NUMBERS

(a) Solve $4 + x = 8$

(b) Solve $-8 + x = 4$

(c) If $x - y = 6$ and $y = 2$, what is the value of x ?

(d) If $2x = y - 20$, find the values of y when $x = 2$ and when $x = -2$

(e) If $3y - \frac{1}{2}x = 6$, write an expression for x in terms of y

(f) Solve $3x - 7 = 2x + 9$

4. QUADRATIC EQUATIONS

(a) How much carpet do I need to cover a floor 5 m long by $2\frac{1}{2}$ m wide?

(b) If the carpet available (a plain one!) only came in a $\frac{3}{4}$ m wide roll, how much would I need off the roll?

(c) What is x times $x + 4$?

(d) What is y times $4y + 3x + 2$?

5. EXPANDING BRACKETS

(a) Does $x(2 + 2x) = 4x^2 + 2x$?

(b) Does $(3x + 2)(x - 1) = 3x^2 + 5x + 2$?

(c) Expand $(3x - 2)(x + 1)$

(d)   Expand $(x - y)(x - y)$

(e)   Expand $(x - y)(x + y)$

(f)   Expand $(-x)(x + 1)(x + 2)$

(g)   Expand $(x + 3)(y^2 + 2)$

(h)   Does $(x - 1)(x^2 + 3x - y) = x^3 + 2x^2 - 3x - xy + y$?

6.   RESTORING BRACKETS

(a)   Express $x^2 + 3x + 2$ in bracket form, thus (   )(   )

(b)   Express $x^2 + x - 2$ in bracket form

(c)   Does $x^2 + 2x - 3 = (x - 1)(x + 3)$?

(d)   If $x^2 + x - 12 = 0$, is it true that $(x - 3)(x + 4) = 0$?

(e)   Solve $x^2 + 7x + 12 = 0$

(f)   Solve $y^2 + \frac{y}{2} - 3 = 0$

7.   GRAPHS AND CO-ORDINATES

(a)   Draw the graphs of the following data on the same diagram

x	-3	-2	-1	0	1	2	3
y	-0.75	-0.25	0.25	0.75	1.25	1.75	2.25

and

x	-1	0	1	2	3
y	4	3	2	1	0

(b)   What are the values of the slopes (or gradients) of the two straight lines?

(c)   What are the co-ordinates of the intersection of the straight lines?

(d)   A motor car enters a motorway. At various times, t, from the entry it is at a distance, s, from the point of entry. From the figures in the table draw a graph and deduce what you can about the way the car is travelling. What is the relationship between s and t? What do you know about the speed of the car?

t (minutes)	0	2	4	6	8	10	12	14	16
s (miles)	0	2	4	6	8	10	12	14	16

(e)   A stone is dropped over the edge of a sheer cliff. The table below gives the distance fallen by the stone at various times from when it was released.

Time (seconds)	0	1	2	3	4	5	6
Distance fallen (metres)	0	5	20	45	80	80	80

Draw a graph of distance fallen against time. What can you say about the relationship between the two? What can be deduced about the cliff?

(f)   The following table shows measurements taken of a car moving along a road. The table shows distance moved from a garage at certain intervals of time.

Time (minutes)	0	10	20	30	40	50	60	70	80	90
Distance (miles)	0	10	20	20	20	40	60	40	20	0

(i)   What is the car's speed during the first 20 minutes?

(ii)   What is happening between 20 and 40 minutes?

(iii)   What does the graph tell us about the journey of the car?

(iv)   What is the total distance moved by the car?

8. THE EQUATION OF A STRAIGHT LINE

A straight line is described by an equation of the form, $y = mx + c$, where m is the gradient or slope, and c is the intercept of the y axis (that is, $y = c$, $x = 0$). What is the equation of line A in 9 (figure 73, p. 143)?

9. FROM EQUATION TO GRAPH

(a)   Draw the graph of $y = -x$

(b)   Draw the graph of $4y = 2x + 3$

(c)   Do the graphs $4y = 2x + 3$ and $y = 0.5x + 0.75$ cross?

10. SOLVING SIMULTANEOUS EQUATIONS GRAPHICALLY

(a)   Find a solution to $y = -x$, and $4y = 2x + 3$ graphically.

(b)   Is there a solution to the following pair of simultaneous equations: $y = -x + 3$ and $y = -x$.

11. SOLVING SIMULTANEOUS EQUATIONS ALGEBRAICALLY

(a)   What value of x satisfies both $y = 0.5x + 0.75$ and $y = 3$?

(b)   Find a solution to problem 10(a) algebraically.

12. RECIPROCALS

What are the reciprocals of :

(a) $\frac{1}{2}$?   (b) 4?

(c)   1?   (d) 0.25?

(e)   −5?   (f) 0.625?

(g)   100?   (h) 10,000?

13. QUADRATIC EQUATIONS

(a)   Draw the graph of $y = -x^2 + 3$

(b)   Draw the graph of $x = -y^2 + 3$

(c)   What are the values of x and y at the crossing points of graphs (a) and (b)?

(d)   Do the graphs of the equations $y = -x^2$ and $y = -x^2 + 3$ cross?

(e)   Draw the graph of $y = -x^2 + x + 3$ and compare it with the graph of $y = -x^2 + 3$.

14. SOLVING QUADRATIC EQUATIONS BY FORMULA

(a)   Solve $-x^2 + 9 = 0$

(b)   Solve $-x^2 + x + 3 = 0$

(c)   Solve $x^2 + 8x + 15 = 0$

(d)   Solve $7x + 22 = 10 - x^2$

15. LINEAR INTERPOLATION

(a)   The following is part of a table of results that were taken during an experiment. Unfortunately some of the values were not noted down at the time. Can you estimate what these values were?

t (mins)	0	1	2	3	4	5	6	7	8
T (°C)	112	100	88	?	64	52	40	?	?

(b)   At what time did a temperature of 70° occur?

(c)   What do you estimate the temperature will be at 10 mins?

(d)   A certain function can be expressed by the following table of values

x	0	1	2	3	4	5	6
y	6	5.5	4.5	3	1.5	0.5	0

(i)   Using linear interpolation between the points $x = 0$ and $x = 3$, estimate the values at $x = 1$ and $x = 2$ and compare them with the values given in the table.

(ii)   Which section of the curve do you think could be best approximated to by linear interpolation?

16. INDICES

(a)   What does 3^2 = ?

(b)   What is $27^{1/3}$?

(c)   What is $2^4 + 3^2$?

(d)   Simplify $a^{1/4} \times a^{1/2}$

(e)   Simplify $a^{1/2} \times a^{-1/2}$

(f)   Evaluate $2^{2.5}$

(g)   Evaluate $\dfrac{3^{0.25}}{2^{1.5}}$

(h)   What does $(2^{3/2})^2$ = ?

17. SCIENTIFIC NOTATION

Express the following numbers in scientific notation:

Example: $2736 = 2.736 \times 10^3$

(a)   52,810

(b)   173,924

(c)   365

(d)   10,078

(e)   6,329,480

(f)   0.0040082

(g)   0.0000002223

(h)   0.174

(i)   0.0602

(j)   1.609

18. LOGARITHMS

What are the logs of:

(a)   2.65

(b)   10.05

(c)   0.00625

(d) + 4.5

(e) + 0.125

What numbers do the following logs represent:

(f)   0.8633

(g) 1.8633

(h) $\bar{1}$.8633

(i) $\bar{3}$.7660

(j)   4.3173

Calculate the following using log tables:

(k)   47.36 x 54.5

(l) 0.256 x 120

(m) 126 ÷ 625

(n) $\dfrac{5.4 \times 6.2}{12.1}$

(o) $(36)^3$

(q) $(36)^{-\frac{1}{3}}$

(p) $(36)^{\frac{1}{3}}$

(s) $(0.036)^{\frac{1}{3}}$

(r) $\dfrac{0.345}{0.00678}$

19. ERRORS AND SIGNIFICANT FIGURES

(a)   Round the following numbers to 4 decimal places:

(i) 4.546848484

(ii)   0.00169124

(iii) −356.7349521

(iv)   −1.25555795

(b)   Express the numbers in the previous exercise to 4 significant figures.

20. TRIANGLES AND TRIGONOMETRIC FUNCTIONS

Are the following statements true?

(a)   An isosceles triangle has all three sides of the same length

(b)   An equilateral triangle has only two of its sides equal

(c)   The sum of the internal angles of a triangle is equal to two right angles

Evaluate the following:

(d)   In a right angled triangle, what is the length of the hypotenuse if the lengths of the other two sides are 5 cm and 12 cm?

(e)   What is the value of sin 30°?

(f)   Does sin 30° = cos 30°?

(g)   Does sin 30° = cos 60°?

(h)   Does sin 45° = cos 45°?

Answers to Exercises

CHAPTER 2

1.	$1\frac{5}{42}$		7.	$2\frac{1}{2}$
2.	$1\frac{4}{33}$		8.	$1\frac{1}{2}$
3.	$8\frac{1}{6}$		9.	$\frac{1}{2}$
4.	$1\frac{7}{12}$		10.	2
5.	$1\frac{3}{4}$		11.	$\frac{2}{3}$
6.	$\frac{1}{15}$		12.	$\frac{5}{6}$

CHAPTER 3

13.	3		25.	−14
14.	−3		26.	−36
15.	1		27.	−63
16.	−7		28.	−63
17.	0		29.	14
18.	1		30.	12
19.	−1		31.	9
20.	−9		32.	−16
21.	13		33.	16
22.	1		34.	−17
23.	−14		35.	1
24.	14		36.	6

37. −6
38. −6
39. 3
40. 3
41. YES
42. YES
43. YES
44. YES
45. YES
46. YES
47. YES
48. YES
49. 1
50. 0
51. (i) 4
    (ii) −6
    (iii) 3
52. (i) 11
    (ii) −3
    (iii) 3¾
    (iv) 9
    (v) −3
    (vi) −4½
53. (i) 14
    (ii) −10
    (iii) −14
54. (i) NO
    (ii) YES
    (iii) 6
    (iv) 6

55. (i) When $x = 3$,
        $y = 9$
        When $x = 0$,
        $y = 6$
        When $x = -5$
        $y = 1$
    (ii) When $y = 0$,
        $x = -6$
        When $y = 7$,
        $x = 1$
        When $y = -1$,
        $x = -7$
56. $y = 10, 9, 8, 7, 6,$
    $5, 4, 3, 2, 1, 0$
57. YES
58. When $x = 3, y = 9$
    When $x = 7, y = 17$
    When $x = 0, y = 3$
59. $x = \dfrac{y + 2}{7}$
60. $x = \dfrac{y}{7} + 2$
61. If you remove the
    brackets you will get
    $7x - 14$ which is the
    same as the second
    equation.
62. They are *all* the same
63. $x = 4$
64. $x = 0$
65. $x = 1½$
66. $x = ⅔$

67.	$x = 5\frac{1}{3}$	69.	$x = -1$
68.	$x = -\frac{2}{3}$	70.	$x = 6$

CHAPTER 5

71. 24 sq.cm or 24 cm^2

72. You could consider
the area to be made up
of 4 strips each 6 cm
long by 1 cm, that is,
the area of each strip
would be 6 sq.cm.
But there are 4 of
them, so the total area
is 4 x (6 sq.cm) =
24 sq.cm

73. $x.y$ sq.cm

74. x^2 sq.cm (i.e. $x.x$)

75. $6x.y$ sq.cm

76. $9x^2$ sq.cm

77. $15xy$ sq.cm

78. $5x(3x + 4)$ sq.cm
or
$15x^2 + 20x$ sq.cm

79. One area is
$(5x)$ x $(3x)$
and the other is
$(5x)$ x 4

80. YES

81. YES

82. $x^2 + 2x$

83. $4x + 2x^2$

CHAPTER 6

84. $3x^2 - x$

85. $3x - 3x^2$

86.

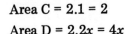

Figure 37

Area A = $x.2x = 2x^2$
Area B = $x.1 = x$
Area C = $2.1 = 2$
Area D = $2.2x = 4x$
∴ Total area =
$2x^2 + x + 2 + 4x$
Or Length = $2x + 1$
    Breadth = $x + 2$
∴ Area =
    $(2x + 1)(x + 2)$

87. YES

88. $(2x + 1)(3x + 2) =$
       $6x^2 + 4x + 3x + 2$

89. $x^2 + 11x + 30$

90. $3x^2 + 19x + 20$

91. $x^2 + 7x + 12$

92. $ac + ad + bc + bd$

93. $x^2 + x - 6$

94. $x^2 - 5x + 6$

95. $2x^2 - 7x + 3$

96. $x^2 - 9$

97. $x^2 - y^2$

CHAPTER 7

98. $(x - 1)(x - 5)$

99. $(x - 2)(x - 3)$

100. $(x + 2)(x + 3)$

101.   This cannot be done

102. $(x - 2)(x + 2)$

103. $(4x + 3)(3x - 2)$

104. $x = 1$

105. $x = -3$

106.   YES

107.   YES, because you
       have not *changed*
       the equation only
       factorised it.

108.   YES

109.   YES

110. $x = 1$ and $x = -3$

111.   YES

112. $(x - 3)(x - 4)$ and
       $x = 3, x = 4$

113. $x = 3$ and $x = -2$

114. $x = -\frac{1}{2}$ and $x = 1$

115. $x = -\frac{1}{3}$ and $x = \frac{5}{2}$

CHAPTER 8

116.   Point A:
       (x co-ordinate is 1,
       y co-ordinate is 1)
       Point B:
       (x co-ordinate is 3
       y co-ordinate is 4)

       Point C:
       (x co-ordinate is 4
       y co-ordinate is 3)
       Point D:
       (x co-ordinate is 5
       y co-ordinate is 2½)

117.

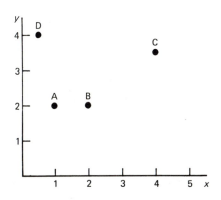

Figure 38

118.   The point would be somewhere on the *y* axis

119.

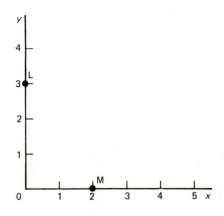

Figure 39

120.   The origin is the point (*x* co-ordinate 0, *y* co-ordinate 0)

121.   A   (2, 2)   B   (4, 3)

      C   (−2, 1)   D   (−1, −2)

      E   (−4, −3)   F   (2, −2)

      G   (5, −2)   H   (5, 0)

      J   (0, 2)   K   (−1, 0)

      L   (0, −5)   P   (4.3, −1.1)

122 and 123.

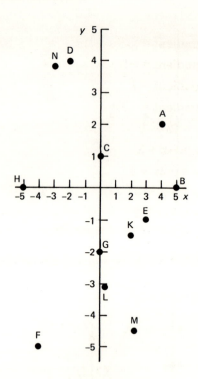

Figure 40

124.   rise = 3

run = 6

$$\therefore \text{gradient} = \frac{\text{rise}}{\text{run}} = \frac{3}{6} = \frac{1}{2}$$

125.   S (2, 6) and T (6, 2)

rise = −4   run = 4

$$\therefore \text{gradient} = \frac{\text{rise}}{\text{run}} = \frac{-4}{4} = -1$$

126.   gradient $\frac{1}{10}$ ft

CHAPTER 9

127.　(i)　gradient = $\frac{1}{3}$

　　　(ii)　gradient = -2

　　　(iii)　gradient = 1

　　　(iv)　gradient = $-\frac{2}{3}$

128.　(i)　$y = 2x + 1$

　　　(ii)　$y = -x + 3$

　　　(iii)　$y = -2x + 4$

　　　(iv)　$y = -x - 3$

　　　(v)　$y = \frac{1}{3}x + 1$

　　　(vi)　$y = x$

　　　(vii)　$y = 2$

CHAPTER 10

129 and 130.

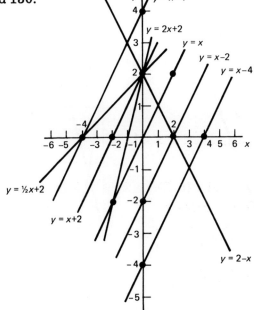

Figure 41

131.

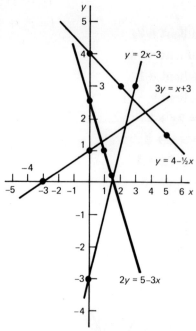

Figure 42

CHAPTER 11

132.

(i)

(ii)

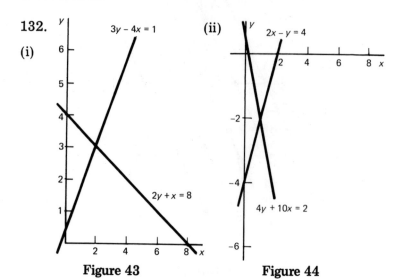

Figure 43   **Figure 44**

(iii)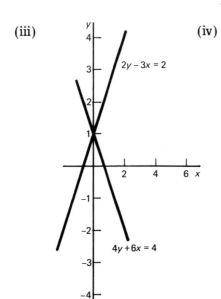

(iv) There is no solution because the lines are parallel.

Figure 45

133. Box 1 : $33 + 8x$
      Box 2 : $33 + 8a$
      Box 3 : $(33 + 8a) \times 2$
      Box 4 : $(33 + 8a) \times 2$
      Box 5 : $(33 \times 2) + (8a \times 2)$
      Box 6 : $66 + 16a$
      Box 7 : 66
      Box 8 : $16a$
      Box 9 : $-65 = 20a$
      Box 10 : $a = -\dfrac{65}{20}$
      Box 11 : $-\dfrac{65}{20}$
      Box 12 : $1 + \dfrac{4 \times 65}{20}$
      Box 13 : $\dfrac{7}{3}$

Box 14 : $33 - \dfrac{8 \times 65}{20}$

Box 15 : $\dfrac{7}{3}$

Box 16 : $\dfrac{7}{3}$

Box 17 : $x = \dfrac{-65}{20}$ and $y = \dfrac{7}{3}$

Box 18 : $2 \times \dfrac{7}{3} - 4 \times \dfrac{65}{20} = 1 = \text{RHS}$

Box 19 : $3 \times \dfrac{7}{3} + 8 \times \dfrac{65}{20} = 33 = \text{RHS}$

Box 20 : $x = \dfrac{-65}{20}$, $y = \dfrac{7}{3}$

134.   (i) $x = 2, y = 3$

(ii) $x = 1, y = -2$

(iii) $x = 0, y = 1$

(iv) $x = 1, y = \frac{1}{2}$

(v) $x = -2, y = 2$

135.

x	1	2	3	4
y	1	½	⅓	¼

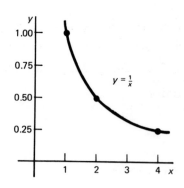

Figure 46

136.

x	¼	½	1	2	4
$\frac{1}{x}$	4	2	1	½	¼

x	−¼	−½	−1	−2	−4
$\frac{1}{x}$	−4	−2	−1	−½	−¼

137.

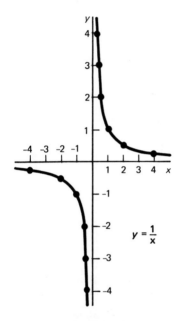

$$y = \frac{1}{x}$$

Figure 47

138. As x becomes very small y becomes very large. Also as y becomes very small x becomes very large. So the graphs go off to infinity along the x and y axes.

139. $y = \frac{1}{0} = $ infinity

CHAPTER 14

140. When $x = 1, y = 1$
When $x = -1, y = 1$
When $x = 3, y = 9$

141.

x	−3	−2	−1	0	1	2	3
y	9	4	1	0	1	4	9

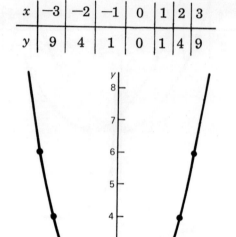

Figure 48

142.   Yes the graph can be extended, but as x becomes
large, y becomes very large indeed, eventually going
off to infinity. This also happens as x becomes very
large but negative.

143.

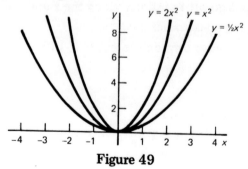

Figure 49

144.

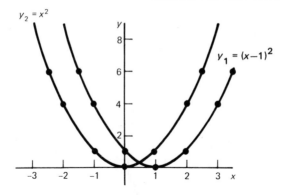

Figure 50

145.

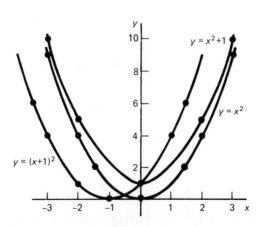

Figure 51

146. Yes, because $y = (x + 1)^2 - 1$
expanded becomes $y = x^2 + 2x$

147. These are shown on the graph of 145

148. $(x + 2)^2 - 5$ expanded becomes
$$x^2 + 4x + 4 - 5 = x^2 + 4x - 1$$

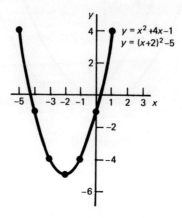

Figure 52

149.

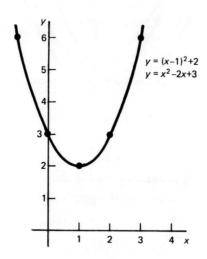

Figure 53

150.   If you expand $y = (x + 1)(x - 3)$ you obtain
        $y = x^2 + x - 3x - 3 = x^2 - 2x - 3$
        When $x = 0$, $y = -3$
        When $x = -1$, $y = 0$

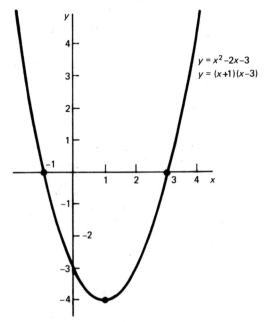

Figure 54

151.

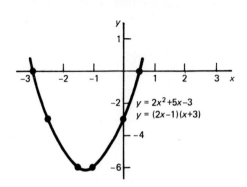

Figure 55

152.

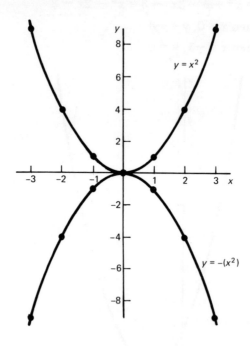

Figure 56

153.   The effect of adding 2 is to move the graph of $y = -x^2$, up the y axis as compared with Exercise 145 where $y = x^2 + 1$ is the graph of $y = x^2$ moved 1 up the y axis.

154.

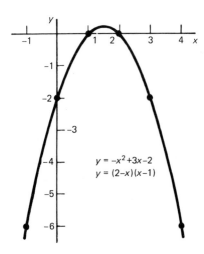

$$y = -x^2 + 3x - 2$$
$$y = (2-x)(x-1)$$

Figure 57

CHAPTER 15

155. (a) $x = -2$ and $x = \frac{1}{2}$

  (b) $x = 2$

  (c) Imaginary roots

156. (a) $(b^2 - 4ac) = 0$

  $\therefore$ root is $= \dfrac{-b}{2a} = \dfrac{1}{3}$

  (b) $(b^2 - 4ac)$ is negative hence the roots are
    imaginary

157. (a) imaginary roots

  (b) $x = -2$ and $x = \frac{1}{2}$

  (c) $x = 2$

  (d) imaginary roots

  (e) $x = \dfrac{7 \pm \sqrt{(5)}}{2}$

  (f) $x = \dfrac{-7 \pm \sqrt{(93)}}{2}$

CHAPTER 16

158.   10 and 14
159.   9, 12 and 15
160.   1100
161.   64.04 °C
162.   46.76 °C
163.   In Exercise 161 the value calculated was 64.04 °C compared with an actual value from the table of 63.90 °C, that is an error of 0.14°C.

In Exercise 162 the error is 0.61°C. So when the time range was much greater in the second case the actual error was also much larger. This is because the curve flattens out as time increases, and therefore you must either take a very close time interval if the temperature is changing quickly or else you can take a large time interval provided the temperature curve is changing slowly.

164.   (i)   70.20 °C
        (ii)   71.52 °C
165.   (i)   0.2995
        (ii)   0.1127
        (iii) 0.6040

166.   When $x = 0.7$,
        $y = 0.2431$
        When $x = 0.65$,
        $y = 0.2587$
        When $x = 1.43$,
        $y = 0.0769$

CHAPTER 17

167. $a^4 \times a^3 = a^7$
        $10^2 \times 10^3 = 10^5$
168.   No, because the
        bases are different

169. 2^5
        2^3
        $2^{1/2}$ or $\sqrt{2}$
170. $a^{7/12}$ a^6
        $a^{3/4}$ a^6

171. 10^{-3}

     Positive

172. They are all equal to 1

173. 2^2

174. 1

175. 81

     64

176. 576

177. 100

     100

     $a^{(2m + 2n)}$

178. (a) 2.3065×10^4

     (b) 9.81665×10^2

     (c) 2.2414×10^4

(d) 2.7316×10^2

(e) 4.129×10^{-2}

(f) 1.336×10^{-1}

(g) 7.51×10^{-4}

(h) 1.00×10^{-6}

179. (a) 2312

     (b) 79.5

     (c) 45900

     (d) 6790000

     (e) 0.0000000229

     (f) 0.000030077

     (g) 0.746

     (h) 4.03

     (i) 0.0352

CHAPTER 19

180. (i)   147.2

     (ii)   2.613

     (iii) 0.1402

     (iv)   4.27

     (v)   0.0000425

     (vi)   4.492

     (vii) 0.4178

     (viii) 0.002016

     (ix)   15.126

CHAPTER 20

181. (a)

     (i)   3.1416

     (ii)   2.7183

     (iii) 22.4591

     (iv)   23.1407

     (v)   65.0515

     (vi)   6.7136

     (vii) 0.0123

     (viii) 176.4329

     (b)

     (i)   3.142

     (ii)   2.718

     (iii) 22.46

     (iv)   23.14

     (v)   65.05

     (vi)   6.714

     (vii) 0.01235

     (viii) 176.4

CHAPTER 21

182. (a)   AB = 10 cm

BC = 5 cm

$\sin \theta = \dfrac{BC}{AB} = \dfrac{5}{10} = \dfrac{1}{2}$

$\therefore \sin \theta = 0.5$

$\theta = 30^{\circ}$

(b)   AB = 10 cm

AC = 5 cm

$\cos \theta = \dfrac{AC}{AB} = \dfrac{5}{10} = \dfrac{1}{2}$

$\cos \theta = 0.5$

$\theta = 60^{\circ}$

(c)   BC = 5 cm

AC = 5 cm

$\tan \theta = \dfrac{5}{5} = 1$

$\theta = 45^{\circ}$

(d)   AC = 1.2941 cm

AB = 5 cm

$\cos \theta = \dfrac{AC}{AB} = 1.2941/5 = 0.2588$

$\theta = 75^{\circ}$

Answers to Revision Chapters 4, 13 and 18

CHAPTER 4

R.1. $x = 2$

R.2. $x = 3$

R.3. $x = -3$

R.4. $x = -15$

R.5. $x = 2$

R.6. $x = -3$

R.7. $x = -\frac{1}{4}$

Rearranging Formulae

1. $N = PD - 2$

2. $b = \dfrac{kv}{k + rt}$

3. $r = \dfrac{C}{2\pi}$

4. $W = \dfrac{P - b}{a}$

5.   (i) $n = \dfrac{2s}{(a + l)}$

     (ii) $l = \dfrac{2s - na}{n}$

6. $P = \dfrac{100A}{(100 + RT)}$

7. $h = \dfrac{S}{2\pi r} - r$

8.   (i) $s = \dfrac{v^2 - u^2}{2a}$

     (ii) $u = \sqrt{(v^2 - 2as)}$

9. $W = \dfrac{LaP}{h - La}$

10. $L(R - r) = 2aE$

     $LR - Lr = 2aE$

     $R = \dfrac{2aE + Lr}{L}$

11. $R^2(Q - bx) = ax - P$

     $R^2 Q - R^2 bx = ax - P$

     $ax + R^2 bx = R^2 Q + P$

     $x = \dfrac{R^2 Q + P}{(a + R^2 b)}$

12. $h = \dfrac{2D^2}{3}$

13. $r = \sqrt{\dfrac{S}{4\pi}}$

14. $T^2 = \dfrac{4\pi^2 I}{MH}$

     $M = \dfrac{4\pi^2 I}{T^2 H}$

15. $u = \sqrt{\left[v^2 - \dfrac{2A}{m}\right]}$

16. $d^2 = a^6 \dfrac{H}{N}$

$H = \dfrac{Nd^2}{a^6}$

17. $\dfrac{4M}{Wd} = 1 - \dfrac{d}{2}$

$l = \dfrac{4M}{Wd} + \dfrac{d}{2}$

18. $Sl = Wdl - \dfrac{Wd^2}{2}$

$l(Wd - S) = \dfrac{Wd^2}{2}$

$l = \dfrac{Wd^2}{2(Wd - S)}$

19. $275H = T\pi Rn - t\pi Rn$

$t = \dfrac{T\pi Rn - 275H}{\pi Rn}$

20. $2H = WR^2 - Wr^2$

$r = \sqrt{\left(\dfrac{WR^2 - 2H}{W}\right)}$

21. (i) $T^2(4 + a^2) = Pbh$

$b = \dfrac{T^2(4 + a^2)}{Ph}$

(ii) $T^2(4 + a^2) = Pbh$

$4T^2 + 4a^2 = Pbh$

$4a^2 = Pbh - 4T^2$

$a = \sqrt{\left(\dfrac{Pbh - 4T^2}{4}\right)}$

22. $3x(a + b) = (a + 2b)h$

$3xa + 3xb = ah + 2bh$

$3xa - ah = 2bh - 3xb$

$a(3x - h) = 2bh - 3xb$

$a = \dfrac{2bh - 3xb}{3x - h}$

23. $v^2 = w^2(a^2 - x^2)$

$v^2 = w^2a^2 - w^2x^2$

$x^2 = \left(\dfrac{w^2a^2 - v^2}{w^2}\right)$

$x = \dfrac{aw - v}{w}$

24. $A^2 = \pi^2 r^4(h^2 + r^2)$

$A^2 = \pi^2 r^4 h^2 + \pi^2 r^6$

$A^2 - \pi^2 r^6 = \pi^2 r^4 h^2$

$h^2 = \dfrac{A^2 - \pi^2 r^6}{\pi^2 r^4}$

$h = \dfrac{A - \pi r^3}{\pi r^2}$

25. (i) $4T = 4H + W^2 l^2$

$H = \dfrac{4T - W^2 l^2}{4}$

(ii) $4T = 4H + W^2 l^2$

$l = \sqrt{\left(\dfrac{4T - 4H}{W^2}\right)}$

26. $\dfrac{uv}{u + v} = \dfrac{f}{2}$

$2uv = uf + vf$

$2uv - uf = vf$

$u(2v - f) = vf$

$u = \dfrac{vf}{(2v - f)}$

27. $9V^2 = \dfrac{S^3}{8\pi}$

$S = \sqrt[3]{72V^2\pi}$

28.
$$A^2(2P^2 - Q^2) = P^2 - 2Q^2$$
$$2A^2P^2 - A^2Q^2 = P^2 - 2Q^2$$
$$P^2(2A^2 - 1) = Q^2(A^2 - 2)$$
$$P^2 = \frac{Q^2(A^2 - 2)}{2A^2 - 1}$$
$$P = Q\sqrt{\left[\frac{A^2 - 2}{2A^2 - 1}\right]}$$

29.
$$A^2 = \pi^2 r^2(h^2 + r^2 + \pi r^2)$$
$$A^2 - \pi^2 r^4 - \pi^3 r^4 = \pi^2 r^2 h^2$$
$$h = \sqrt{\left[\frac{(A^2 - \pi^2 r^4 - \pi^3 r^4)}{\pi^2 r^2}\right]}$$

30.
$$r^2 = \frac{f^2}{4} + \frac{f^2}{4} + q^2$$
$$q^2 = r^2 - 2\left(\frac{f^2}{4}\right)$$
$$q = r - \frac{f}{\sqrt{2}}$$

31.
$$e(LT - lt) = L - l$$
$$elt - l = eLT - L$$
$$l = \frac{L(eT - 1)}{et - 1}$$

32.
$$2HfE = 2Hg - V^2$$
$$H(2g - 2fE) = V^2$$
$$H = \frac{V^2}{2(g - fE)}$$

33.
$$V^2 = gd\left(1 + \frac{3h}{d}\right)$$
$$V^2 = gd + \frac{3hgd}{d}$$
$$V^2 d = gd^2 + 3ghd$$
$$V^2 = gd + 3gh$$
$$d = \frac{V^2 - 3gh}{g}$$

34.
$$T^2gh = 4\pi^2(h^2 + k^2)$$
$$T^2gh - 4\pi^2h^2 = 4\pi^2k^2$$
$$k^2 = \frac{T^2gh - 4\pi^2h^2}{4\pi^2}$$
$$k = \frac{T^2gh}{4\pi^2} - h^2$$
$$k = \frac{T}{2\pi}\sqrt{(gh)} - h$$

35.
$$C(a^2 + b^2) = a^2 b^2 2g$$
$$a^2C + Cb^2 = a^2 b^2 2g$$
$$a^2(b^2 2g - C) = \frac{Cb^2}{2}$$
$$a = \sqrt{\left[\frac{Cb^2}{2b^2g - C}\right]}$$

36.
$$2gE = mv^2 - mu^2$$
$$u = \sqrt{\left[\frac{mv^2 - 2gE}{m}\right]}$$

37.
$$2mc + 2c = 3km - 6k$$
$$m(3k - 2c) = 2c - 6k$$
$$m = \frac{2c - 6k}{3k - 2c}$$

38. $D^2 = \dfrac{2v^2 d}{g} + \dfrac{d}{4} - \dfrac{d}{2}$

$8gD^2 = 16v^2 d + 2gd - 4gd$

$8gD^2 = 2d(8v^2 + g - 2g)$

$8gD^2 = 2d(8v^2 - g)$

$d = \dfrac{4gD^2}{8v^2 - g}$

39. $Ke^2(1 + kt) = t - k$

$Ke^2 + kKt = t - k$

$k(Kt - 1) = t - Ke^2$

$k = \dfrac{t - Ke^2}{Kt - 1}$

40. $32xT - 32xW = Wv^2$

$W(v^2 + 32x) = 32xT$

$W = \dfrac{32xT}{v^2 + 32x}$

CHAPTER 13

I   Basic Arithmetic

1.   (i) 31 (ii) 52 (iii) 39 (iv) 13 (v) 12

2.   (i) F   (ii) T   (iii) F   (iv) T   (v) T
     (iv) T   (vii) F   (viii) T   (ix) T   (x) T

II   Basic Algebra

3   *Formulae*

(i) $S = \dfrac{t}{60} x$ (ii) $\dfrac{x}{a} \times b$ (iii) $8, -30$

(iv) $2.5, 1\dfrac{5}{24}$ (v) $A = 176, I = 48, K = 6250$

(vi) $t = 18.84$ (vii) $V = 208.75$

4. *Algebraic Simplification*

   (i) $18x$ (ii) $2x$ (iii) $-5a$ (iv) ab (v) $14xy$

   (vi) $x - 5y + 9z$ (vii) $12ab$ (viii) $-xy$

   (ix) $60abcd$ (x) $\left(\dfrac{a}{b}\right)$ (xi) $2b$ (xii) $-2a^2b$

   (xiii) $\dfrac{7}{3}ab$ (xiv) $9pq^2$ (xv) $-21a^3b$

   (xvi) $9x + 6y$ (xvii) $-a-b$ (xviii) $-9xy -12y$

   (xix) $2m - 6m^2 + 4mn$ (xx) $14 - 2a$

   (xxi) $x - 17$ (xxii) $7x + y$ (xxiii) $16 - 17x$

   (xxiv) $3a - 9b$ (xxv) $-x^3 + 18x^2 - 9x + 15$

5. *Brackets*

   (i) $x^2 + 3x + 2$ (xi) $6x^2 - x - 15$

   (ii) $2x^2 + 11x + 15$ (xii) $4p^2 - q^2$

   (iii) $10x^2 + 17x + 3$ (xiii) $25a^2 - 49$

   (iv) $x^2 - 4x + 3$ (xiv) $6x^2 - xy - 12y^2$

   (v) $2x^2 - 9x + 4$ (xv) $4x^2 + 12x + 9$

   (vi) $4x^2 - 33x + 8$ (xvi) $x^2 - 2x + 1$

   (vii) $6x^2 - 17x + 5$ (xvii) $4a^2 + 12ab + 9b^2$

   (viii) $x^2 + 2x - 3$ (xviii) $a^2 - 2ab + b^2$

   (ix) $2x^2 + x - 10$ (xix) $9x^2 - 24xy + 16y^2$

   (x) $6x^2 + x - 15$ (xx) $x^3 - 2x^2y + xy^2$

6. *Miscellaneous Problems*

   (i) T (ii) F (iii) F (iv) T (v) T

   (vi) T (vii) F (viii) T (ix) F (x) F

III   **Simple Equations**

7. (i) T (ii) F (iii) F (iv) T (v) F (vi) F (vii) T

   (viii) F (ix) T (x) F (xi) F (xii) F

8. (i) a (ii) b (iii) b (iv) a

IV   Simultaneous Equations

9.   (i) a (ii) b (iii) b (iv) a and b (v) b

Graphs, including Solutions of equations

1.

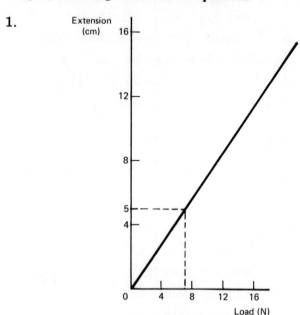

Figure 58

Interpolation

For an extension of 5 cm a load of 7 N is required.

Extrapolation

(i)   Load of 20 N produces extension of 13.9 cm.

(ii)   Hence 100 N might produce 69.5 cm.

However, extrapolation is rather dangerous:

(i)   There is no guarantee that the linear relationship
        (which is Hookes' Law) holds beyond a load of 12 N.
        The spring might fracture or become plastic.

(ii)   Any errors incurred in drawing the line through the
        points will be magnified by extrapolation.

2.

−1	−5
0	−2
1	1
2	4
3	7
4	10

Note from (1) and (ii)
gradient = 3
intercept on $y = -2$
Equation of line is
$y = 3x - 2$
That is, $y = $ (gradient)
$x - $ (intercept on y)

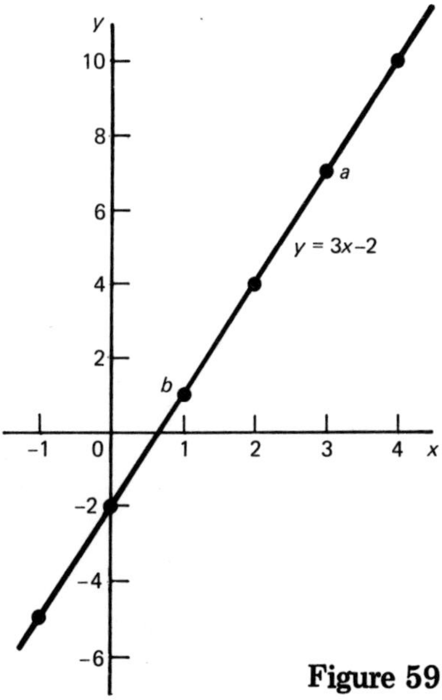

Figure 59

(i) a and b are any two points on the
    straight line
    gradient = $\dfrac{(y)a - (y)b}{(x)a - (x)b}$

$= \dfrac{7 - 1}{3 - 1}$

$= \underline{3}$

(ii) When $y = 0$, $x = \dfrac{2}{3}$

(iii) When $x = 0$, $y = -2$

3.

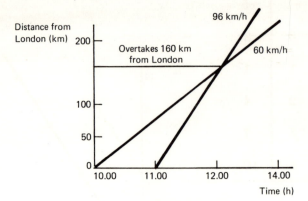

Figure 60

4.

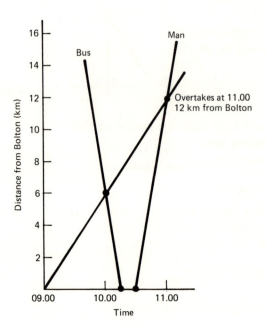

Figure 61

5.　　$P = 3r + 5$

r	0	1	2	3
$P = 3r + 5$	5	8	11	14

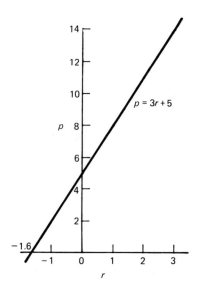

Figure 62

From graph:

solution of $3r + 5 = 0$ occurs when $P = 0$ on the graph, that is at $r = 1.6$

6.   Let $y = 3(x - 2) - 6$

        $= 3x - 12$

x	0	1	2	3	4	5	6
$y = 3x - 12$	−12	−9	−6	−3	0	3	6

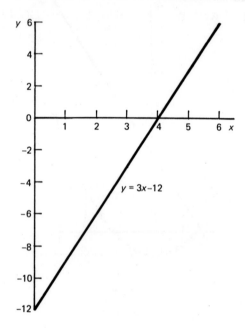

Figure 63

From the graph:
solution of $3x - 12 = 0$ occurs when $y = 0$ on the
graph, that is at $x = 4$

7. $y = x^2 + 9x + 20$

x	−8	−7	−6	−5	−4	−3	−2	−1	0	1	1
x^2	64	49	36	25	16	9	4	1	0	1	4
$9x$	−72	−63	−54	−45	−36	−27	−18	−9	0	9	18
y	12	6	2	0	0	2	6	12	20	30	42

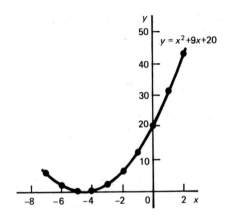

Figure 64

Solutions of $x + 9x + 20 = 0$
occur when $y = 0$ on the graph that is, at

$$x_1 = -4$$

$$x_2 = -5$$

There are two solutions for any quadratic. In this case

$$(x - x_1)(x_2 - x) = (x + 4)(x + 5)$$

$$= x^2 + 9x + 20$$

8.

x	0	1	2	3	4	5	6	7	8
x^2	0	1	4	9	16	25	36	49	64
$-8x$	0	-8	-16	-24	-32	-40	-48	-56	-64
y	32	25	20	17	16	17	20	25	32

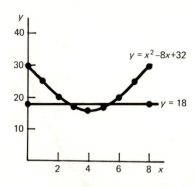

Figure 65

When $y = 18$, $x_1 = 2.6$ or $x_2 = 5.4$

The intercepts of $y = 18$

and $y = x - 8x + 32$

that is, $x^2 - 8x + 32 = 18$

∴ $x^2 - 8x + 14 = 0$

gives solutions for this final expression. Hence to confirm

$$(x - 2.59)(x - 5.41) = x^2 - 8x + 14)$$

9.   Equation 1 $3x + 2y = 12$

$\therefore 2y = -3x + 12$

$\therefore\ y = -\dfrac{3x}{2} + 6$

x	0	2	4
y	6	3	0

Equation 2 $4x - 3y = -1$

$\therefore 3y = 4x + 1$

$\therefore\ y = \dfrac{4x}{3} + \dfrac{1}{3}$

x	0	2	4
y	⅓	3	4⅓

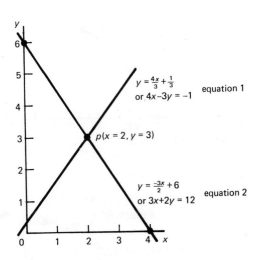

Figure 66

Solution of the simultaneous equations is at P $\therefore$ <u>$x = 2, y = 3$</u>

10. $y = x^2 - x + 7$
        $y = x + 8$

x	-2	-1	0	1	2	3
$y = x^2 - x + 7$	13	9	7	7	9	13
$y = x + 8$	6		8		10	

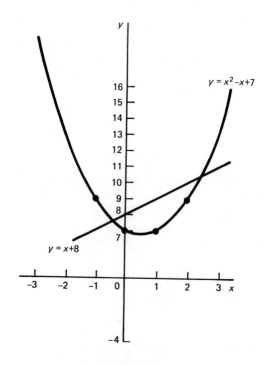

Figure 67

Points of intersection occur when $x^2 - x + 7 = x + 8$ which gives the caption $x^2 - 2x - 1 = 0$ which has roots of $x = 1 \pm \sqrt{(2)}$ that is, $x = 2.41$ or -0.41

11. $y = x^2$

$y = x + 1$

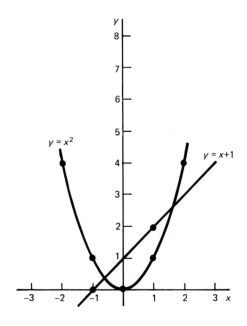

Figure 68

Intersections occur when $x^2 = x + 1$ that is, $x^2 - x - 1 = 0$
which has solutions $x = \dfrac{1}{2} \pm \sqrt{\left(\dfrac{5}{2}\right)}$

$= 1.62$ or -0.62

from the graph $x = 1.65$ and -0.62

CHAPTER 18

1.   100; 10,000; 200; 512; 1536; 26,000,000;
     39,500,000,000; 0.000000000395; 1, 100.

2. 2.56×10^2; 7.295×10^1; 1×10^6 approx,
     1.25265629×10^5; 1.09×10^{-3}; 1.25×10^0;
     1.0×10^{12}

3.   Expressing long numbers simply: e.g. last three
     examples in 1. Very small numbers. for example
     question 1 again.

4. 10^{10}; 10^7, 2.75×10^2, 2^{21}; 2^{21};

     $a^m \times a^n = a^{m+n}$; $\dfrac{a^m}{a^n} = a^{m-n}$; $(a^m)^n = a^{m/n}$;
     $(a^m)^1 = a^{m/n}$.

5. 10^{-1}; 10^5; 5, 5×10^{13}; 2^9; 2^{16}.

6.   10; $\dfrac{1}{10}$; 10; $\dfrac{1}{100}$; 2; $\dfrac{1}{2}$; $\dfrac{1}{2}$; $\dfrac{1}{2}$; 2; 2.

7. 2^{15}; 1,000,000; 10; 27.

Answers to General Revision Chapter 22

1. (a) $3\frac{1}{8}$ (b) $1\frac{3}{8}$
    (c) YES   (d) $\frac{5}{6}$
    (e) $15\frac{3}{4}$ (f) 5
    (g) $1\frac{1}{3}$

2. (a) -1 (b) -11
    (c) 13   (d) 15
    (e) -1 (f) 8
    (g) -8 (h) -8
    (i) -6 (j) 6
    (k) -6 (l) $1\frac{1}{2}$
    (m) $-1\frac{1}{2}$ (n) $-1\frac{1}{2}$
    (o) $1\frac{1}{2}$ (p) YES
    (q) NO

3. (a) $x = 4$ (b) $x = 12$
    (c) $x = 8$ (d) When $x = 2, y = 24$, and
                             when $x = -2$, $y = 16$
    (e) $x = 6y - 12$ (f) $x = 16$

4. (a) $12\frac{1}{2}$ square   (b) $16\frac{2}{3}$ metres
         metres
    (c) $x^2 + 4x$ (d) $4y^2 + 3xy + 2y$

5. (a)   NO   (b)   NO
    (c) $3x^2 + x - 2$ (d) $x^2 - 2xy + y^2$
    (e) $x^2 - y^2$ (f) $-x^3 - 3x^2 - 2x$
    (g) $3y^2 + xy^2 + 2x + 6$
    (h)   YES

6. (a) $(x + 1)(x + 2)$ (b) $(x - 1)(x + 2)$
    (c)   YES   (d)   YES
    (e) $x = -3$, and   (f) $y = 3/2$, and
          $x = -4$ $y = -2$

7(a)

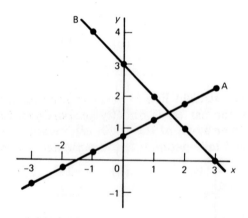

Figure 69

(b)   Slope of A is 0.5 and B is -1
(c) $x = 1.5$, $y = 1.5$ or $(1.5, 1.5)$

(d)

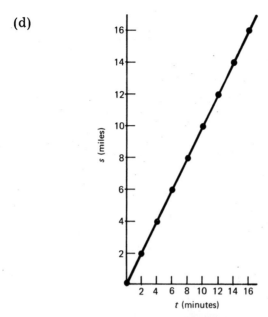

Figure 70

The graph is a straight line so you can say that the distance travelled by the car (*s*) is directly proportional to the time (*t*) elapsed. The slope of the graph is a measure of the car's speed, that is 1 mile per minute (60 mph) and is constant.

(e)

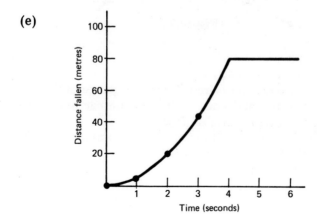

Figure 71

You will see that at first it is not a straight line, but after $t = 4$ seconds it is horizontal. You should be able to deduce from this that the cliff is 80 m high! While the stone is falling the distance travelled is *not* proportional to the time elapsed. In each successive period of 1 second the stone falls a greater distance than before. It is *accelerating* under the force of gravity. If you have the time and the interest you could see what happens if you plot s against t^2 for $t = 1$ to 4 seconds.

(f)

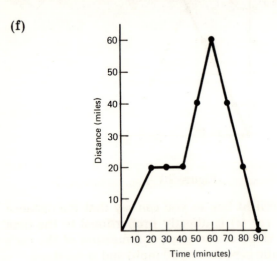

Figure 72

(i)   60 mph
(ii)   The car is stationary.
(iii)   The distance measured is *from* a garage and as the graph starts and ends at 0 then the car must come back to where it started from.
(iv)   60 miles out and 60 miles back, a total of 120 miles.

8. $y = 0.3x + 0.5$

9. (a) and (b)

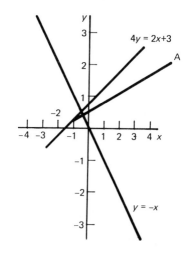

Figure 73

(c) The two equations are the same so it could be said that the lines cross at every point, or not at all!

10. (a)

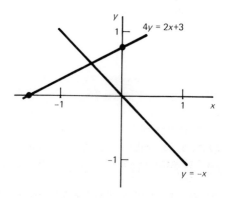

Figure 74

The solution is where the graphs cross, that is, $x = -0.5$, $y = 0.5$.

(b) No, the curve of $y = -x + 3$ is a line *parallel* to $y = -x$ but moved up the y axis by 3.

11.   (a) $x = 4.5$

(b) $x = -0.5, y = 0.5$

12.   (a)   2   (b)   ¼ or 0.25

(c)   1   (d)   4

(e)   −0.2 or −1/5   (f)   8/5 or 1.6

(g)   0.01   (h)   0.0001

13. (a) and (b)

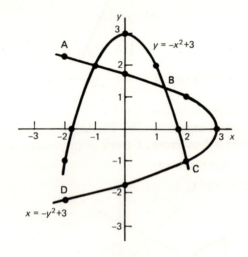

Figure 75

(c)   Crossing points approximately:

A. (1.3, 1.3)   B.   (−1, 2)

C. (2, −1)   D.   (−2.25, −2.25)

(d)   These graphs will not cross since $y = -x^2 + 3$ is the same shape as $y = -x^2$ but moved up the y axis by 3 units.

(e)

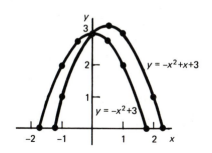

Figure 76

14.   (a) $x = \pm 3$ (b) $x = \frac{1}{2} \pm \dfrac{\sqrt{13}}{2}$

     (c) $x = -5$ or -3 (d) $x = -3$ or -4

15. (a)   At t = 3 minutes, T = 76 °C
           t = 7 minutes, T = 28 °C
           t = 8 minutes, T = 16 °C

     (b)   t = 3.5 minutes

     (c)   T = −8 °C

     (d)

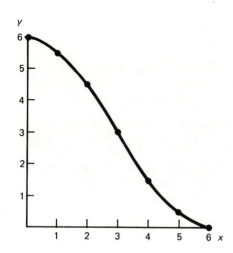

Figure 77

(i) At $x = 1$ $y = 5$ c.f. $y = 5.5$ given on table and at $x = 2$ $y = 4$ c.f. $y = 4.5$ given on table.

(ii) The curve from $x = 2$ to $x = 4$ is the best part to approximate by a straight line. On the rest of the curve its gradient is changing quickly so errors would be introduced if linear interpolation was used.

16. (a) 9   (b) 3
    (c) 25   (d) $a^{3/4}$
    (e) $a^0 = 1$ (f) 5.65685
    (g) 0.4653   (h) 8

17. (a) 5.2810×10^4 (b) 1.73924×10^5
    (c) 3.65×10^2 (d) 1.0078×10^4
    (e) 6.329480×10^6 (f) 4.0082×10^{-3}
    (g) 2.223×10^{-7} (h) 1.74×10^{-1}
    (i) 6.02×10^{-2} (j) 1.609×10^0

18. (a) 0.42325   (b) 1.00212
    (c) $\bar{3}.79588$ (d) 0.65321
    (e) $\bar{1}.09691$ (f) 7.2997
    (g) 72.997   (h) 0.72997
    (i) 0.0058345 (j) 2076.3
    (k) 2581.1   (l) 30.72
    (m) 0.2016   (n) 2.7669
    (o) 46656   (p) 3.3019
    (q) 0.30285   (r) 50.885
    (s) 0.3302

19. (a) (i) 4.5468   (ii) 0.0017
        (iii) −356.7350 (iv) −1.2556
    (b) (i) 4.547   (ii) 0.001691
        (iii) −356.7   (iv) −1.256

20.	(a)	False	(b)	False, in fact all *three* sides are equal in length
	(c)	True	(d)	13 cm
	(e)	0.5	(f)	NO
	(g)	YES	(h)	YES